Explosion Shock Waves and High Strain Rate Phenomena

Conference on Explosion Shock Waves and High Strain Rate Phenomena, ESHP 2019, Puducherry, India 19-21 March, 2019

Editors
K. Hokamoto[1] and K. Raghukandan[2]

[1]Institute of Pulsed Power Science, Kumamoto University, Kumamoto, Japan
[2]Department of Manufacturing Engineering, Annamalai University, Annamalainagar, Tamilnadu, India

Peer review statement

All papers published in this volume of "Materials Research Proceedings" have been peer reviewed. The process of peer review was initiated and overseen by the above proceedings editors. All reviews were conducted by expert referees in accordance to Materials Research Forum LLC high standards.

Published under License by **Materials Research Forum LLC**
Millersville, PA 17551, USA

Published as part of the proceedings series
Materials Research Proceedings
Volume 13 (2019)

ISSN 2474-3941 (Print)
ISSN 2474-395X (Online)

ISBN 978-1-64490-032-1 (Print)
ISBN 978-1-64490-033-8 (eBook)

This book contains information obtained from authentic and highly regarded sources. Reasonable efforts have been made to publish reliable data and information, but the author and publisher cannot assume responsibility for the validity of all materials or the consequences of their use. The authors and publishers have attempted to trace the copyright holders of all material reproduced in this publication and apologize to copyright holders if permission to publish in this form has not been obtained. If any copyright material has not been acknowledged please write and let us know so we may rectify in any future reprint.

Distributed worldwide by

Materials Research Forum LLC
105 Springdale Lane
Millersville, PA 17551
USA
http://www.mrforum.com

Manufactured in the United State of America
10 9 8 7 6 5 4 3 2 1

Table of Contents

Preface

After the successful completion of previous five Explosion, Shock wave and High strain-rate Phenomenon (ESHP) symposiums, the sixth edition was held in the Hotel Ocean Spray, Puducherry, India from 19 to 21 March 2019. The prime objective of this symposium is to discuss the recent progress in the field of explosion, shock wave and high strain rate phenomenon and to establish a network of researchers in the field globally. More than 75 researchers and scientists from different parts of the world viz., Greece, Japan, Singapore, Slovenia, China, USA, England, Iran, Armenia, Bangladesh and India have participated. In order to motivate young researchers, young researcher awards were presented during ESHP 2019. Based on the results of various researchers, a collection of articles was prepared and presented in this edition of Materials Research Proceedings.

We thank all the authors, participants, reviewers, sponsors and all other people for their contribution. We also thank the publisher, Materials Research Forum LLC, of this proceeding volume for their support in publishing these proceedings in a good manner. The next edition of the ESHP symposium will be held in Slovenia.

K. Hokamoto
K. Raghukandan
Co-Chairs
ESHP 2019

Committees

Chairs
Kazuyuki Hokamoto, Kumamoto University, Japan
K. Raghukandan, Annamalai University, India

International Scientific Committee
- Balagansky, I., Novosibirsk State Technical University, Russia
- Ben-dor, G., Ben-Gurion University of the Negev, Israel
- Buchar, J., Mendel University of Agriculture and Forestry, Czecoslovakia
- Chen, P.W., Beijing Institute of Technology, P.R. China
- Deribas, A., Inst. Struct. Macrokinetics& Mater. Sci., RAS, Russia
- Hasebe, T., Kobe University, Japan
- Hoon, H.H., Nanyang Technological University, Singapore
- Kubota, S., National Institute of Industrial Science and Technology, Japan
- Li, X., Dalian University of Technology, P.R.China
- Liu, Z., Beijing Institute of Technology, P.R. China
- Mamalis, A.G., National Center for Scientific Research, Greece
- Moatamedi,M., University of Manchester, UK
- Pruemmer, R., University of Karlsruhe, Germany
- Ren, Z., University of Maribor, Slovenia
- Rittel, D., Israel Institute of Technology, Israel
- Sucesca, M., University of Zagreb, Croatia
- Shin, H.S., Andong National University, Korea
- Shim Victor, P.W., National University of Singapore, Singapore
- Thadhani, N., Georgia Institute of Technology, USA
- Wang, C., Beijing Institute of Technology, P.R. China
- Wang, G., China Academy of Engineering Physics, P.R. China
- Wang, J., Nanjing University of Science and Technology, P.R. China
- Yamashita, M., Gifu University, Japan
- Yoh, J., Seoul National University, Korea
- Zhou, M., Georgia Institute of Technology, USA

Steering Committee
- Siva Kumar, K., Defence Metallurgical Research Laboratory, India
- Dinesh Kumar Balkishan Pal, TBRL India
- Murugan, G., Annamalai University, India
- Saravanan, S., Annamalai University, India

Organizing Committee
- Mahadevan, K., Pondicherry Engineering College, India
- Rajesh Chandra, IDL Explosives, India
- Sathyanarayanan, Alva's Institute of Engineering and Technology, India
- Tamilchelvan, P., Galgotias University, India

Sponsors

- Institute of Pulsed Power Science, Kumamoto University
- Section of Explosion and Impulsive Processing, Japan Explosive Society
- Committee of the High-Energy-Rate Forming, Japan Society for Technology of Plasticity
- State Key Laboratory of Explosion Science and Technology, Beijing Institute of Technology
- Explosion and Safety Division, China Ordinance Society
- High Energy Materials Society of India, Chandigarh-Delhi Chapter
- SAARC Cases
- City Union Bank Limited, India
- IDL Explosives, Hyderabad
- Sakthi Medicare Centre, Chidambaram

Explosion Shock Waves and High Strain Rate Phenomena   Materials Research Forum LLC
Materials Research Proceedings **13** (2019) 1-6   https://doi.org/10.21741/9781644900338-1

The Combustion and Transition to Detonation of High Pressure Flammable Gas in Closed Spaces Linked with a Narrow Path

Yusaku Kusuhara[1*], Kazuhito Fujiwara[1], Fumiko Kawashima, Shingo Maeda[1], Ryo Nanba[1]

[1]Department of Mechanical System Engineering, Kumamoto University2-39-1 Kurokami, Chuo-Ku, Kumamoto 860-8555 Japan

*183d8457@st.kumamoto.u.ac.jp

Keywords: Combustion, High Pressure Flammable Gas, XiFoam

Abstract. When flammable gases confined or compressed in closed space such as metal cases or shells accidentally combusted, the deflagration could be generated and building up to detonation might cause intensive explosion. High energy density has been pursued in some industrial products or in some manufacturing processes, while the risk of troubles is increasing. Generally the combustion transitions to detonation in highly turbulent flows and takes some buildup time or propagation length. But in the complicated and closed space geometry such as the structure of compressors there are many interactions among compressive wave and rigid surface, and then the transition to detonation frequently has been observed. The product design considering the transition phenomena and reducing the risk of explosions is required in high energy fields.

In this study detonations of flammable gas in the high pressure vessel that has spaces linked with narrow curved path were observed and simulated numerically. A high speed camera was used to observe the flame, and the history data were acquired from pressure gauges. In the simulation, XiFoam mounted in Open FOAM was used as the base code. From the visual comparison between the results of the experiment and the simulation, it was shown that turbulent burning velocity suddenly increases and the pressure exceeds a certain value when combustion transition to detonation. These criteria is useful for the design of interior structure of high pressure facilities.

1. Introduction

Recently, compressors are used for various things such as air conditioners and refrigerators. The use of flammable gas as a refrigerant of the compressor is increasing. However, when oxygen and air continue to be sucked into the compressor due to human error, high temperature and high pressure air-fuel mixture is formed inside the compressor. It has been confirmed that the air-fuel mixture spontaneously ignites and a combustion explosion occurs, leading to rupture of the compressor. Therefore, it is necessary to carry out risk assessment on flammable gas.

The form of flame propagation has combustion and detonation. The distinction between them is very clear, and when the combustion wave propagates through the mixture, it suddenly transits to detonation. The combustion is characterized by a slight pressure change, while the detonation pressure rises sharply. Even when combustion occurs in the compressor, the combustion is transitioning to detonation. Therefore, a sudden rise in pressure occurs and it is considered that the compressor has ruptured. It seems that the pressure at the time of burning and the structure of the compressor are influencing the occurrence of detonation.

Therefore, in this study, we investigate the combustion of the premixed gas in the sealed high pressure container, assuming that the combustible gas burned in the compressor. As a survey method, a method of actually performing a combustion experiment and measuring observation and pressure, and a method of reproducing combustion by combustion analysis are used. This time we will use a model that simplifies the shape of the compressor and conduct experiments that will be the basis of future combustion experiments and combustion analysis.

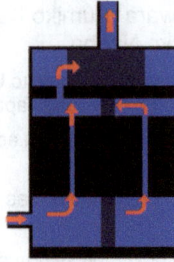

Fig. 1.1 Schematic of compressor

2. Combustion experiment

In order to reproduce the structure of the compressor simply, one divider was installed in the rectangular space one above the other to form a two step slit structure. The position of the two-step slit can be changed. The experimental apparatus and flow path are shown in Fig.2.1 and Fig.2.2.In order to observe combustion, acrylic is attached and visualized on the front of the equipment, and photographing is performed with a high speed camera. A pressure sensor is installed in the upper part of the flow path device, and the sensor position and ignition position are shown in Fig.2.2.

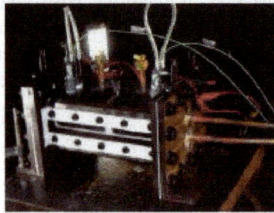

Fig.2.1 Experimental device

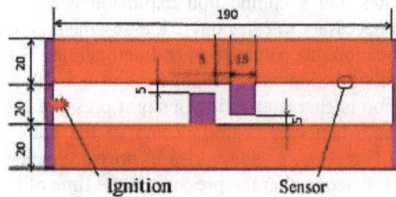

Fig.2.2 Combustion channel and position of sensors

After filling with a mixture of propane and air (equivalence ratio: 1), ignite and burn. As an ignition method, an explosion phenomenon caused by applying an impact high current to metal thin wires is used. As the experimental conditions, experiments are carried out with a total of 4 patterns consisting of two patterns with initial pressure of 1.2 MPa and 0.6 MPa, and two patterns with two slit positions shifted by 40 mm from the center and the center to the ignition side. Fig.2.3 shows the image taken by the high-speed camera for each condition.

Table 2.1 Experiment condition

No.	Initial pressure[MPa]	Slit position
1	1.2	Center
2	1.2	40mm to left from center
3	0.6	Center
4	0.6	40mm to left from center

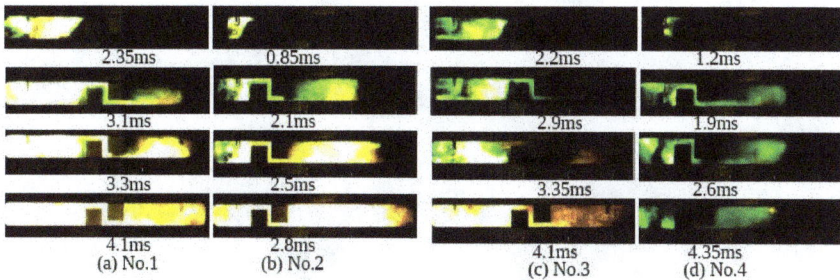

Fig.2.3 Picture of combustion

In Nos. 1 to 3, strong combustion is occurring around the time the flame starts to spread through the two-stage slit. It is considered that transition to detonation occurs at this timing. On the other hand, in No. 4, combustion of almost constant strength is propagated and detonation has not occurred. Also, since the flame front is shaking back and forth as the combustion propagates, it was found that shock waves were generated during combustion and the reflection was repeated in the space.

Next, a graph comparing the pressure under each condition is shown in Fig.3.4.

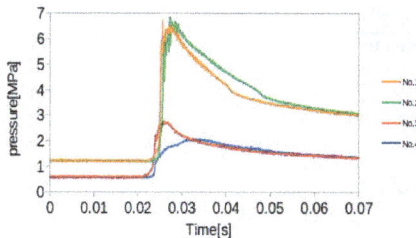

Fig.2.4 Comparison of pressure

Nos. 1 and 2 show almost the same waveform, and when detonation occurs, it is understood that the position of the slit has almost no effect on pressure. From the comparison of No. 3 and

3

No. 4, it can be seen that when the detonation is occurring, the pressure rising speed increases considerably and the maximum pressure also increases. In No. 3, the time until combustion passes through the slit is long, and the pressure increases as the number of reflections of the shock wave increases in the space behind the slit. It seems that it may be easy to transition to detonation by combustion propagation to that space. From this, it is considered that safety is higher when the pressure at ignition is lower, the position of the slit closer to the ignition side.

3. Combustion analysis

For combustion analysis, OpenFOAM[1] which is a numerical fluid dynamics tool box is used, and XiFoam is used as a solver. XiFoam is a solver for premixed combustion considering compressibility / turbulence model[2]

The model produced what reproduced the combustible space in the combustion experiment. Four patterns of conditions similar to combustion experiments were analyzed. The model used is shown in Fig. 3.1, and the analysis conditions are shown in Table 3.1. The pressure measurement is performed at the same position as the combustion experiment.

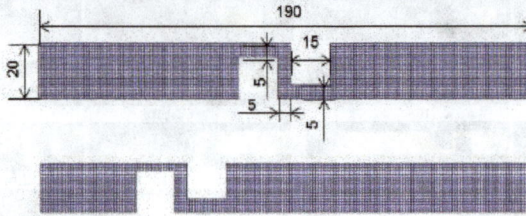

Fig.3.1 Analysis model

Table3.1 Analysis condition

Use solver	XiFoam
Mesh spacing	$\Delta x = \Delta y = \Delta z = 1$[mm]
Gas used	propane
Initial pressure	0.6 or 1.2[MPa]
Ignition position	0, 10, 0.5
Calculation time interval	1 e -6 [s]
Number of cells	3350
Equivalent ratio	1
Temperature	300[K]
Ignition source diameter	5[mm]

Fig.3.2 shows pressure data obtained by combustion analysis by XiFoam.

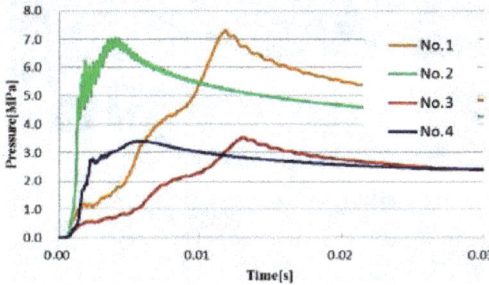

Fig.3.2 Comparison of pressure

According to Fig.3.2, in the analysis by XiFoam, there is no difference in the maximum pressure due to the position of the slit, and the maximum pressure largely depends on the initial pressure. Compared with the pressure in Fig. 2.4, when the slit position is in the middle, the pressure rise rate is smaller in the analysis.

Since the combustion that causes the transition to detonation can not be reproduced by XiFoam, investigate the transition condition and modify the solver accordingly so that the detonation can be reproduced. The turbulent combustion speed rapidly increased at any of the conditions No. 1 to No. 3, as the combustion began to spread through the slit, and in addition, the pressure increased to a constant pressure in addition all right. The temperature distribution at No.2 in that case is shown in Fig.3.3, and the turbulent burning velocity distribution is shown in Fig.3.4.

Because conditions for transition to detonation were obtained, the condition that the turbulent burning speed suddenly rises and the pressure becomes 100% when the pressure exceeds the constant pressure is incorporated into the program of the XiFoam solver, Reproduce the transition. The values of these conditions were determined to be consistent with the experimental results.

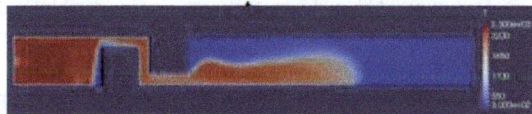

Fig.3.3 Temperature profile

Fig.3.4 Turbulent burning velocity profile

4. Comparison between combustion experiment and combustion analysis

Analysis using an improved solver gave similar data in Nos. 1 to 3, so the typical temperature distribution analyzed under No. 2 is shown in Fig. 4.1, the pressure data is shown in Fig. It is shown in 4.2.

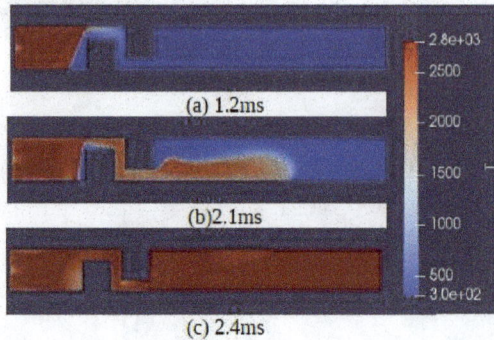

(a) 1.2ms

(b)2.1ms

(c) 2.4ms

Fig.4.1 Temperature profile

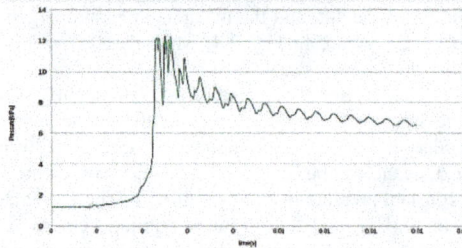

Fig.4.2 Pressure history

As you can see in Fig. 4.1, the flame starts to spread through the slit, the combustion is rapidly strengthened and the flame spreads throughout. The pressure rapidly rises after transition to detonation, rising by 20 MPa from before the transition and the maximum pressure is about 20 times the initial pressure. Although the pressure is considerably high, it is thought that the reproducibility of detonation is high with respect to the propagation of combustion.

Conclusion

From the visual comparison between the results of the experiment and the simulation, it was shown that turbulent burning velocity suddenly increases and the pressure exceeds a certain value when combustion transition to detonation.

We developed a solver which simulates the conditions leading to combustion by detonation in a model simulating a compressor in a simplified manner and improves the reproducibility of the detonation change.

References

[1] PENGUINTIS Open FOAM information
(http://www.geocities.jp/penguinitis2002/study/OpenFOAM/ OpenFOAM-info.html)

[2] Prototype Open FOAM Standard Solver List (http://dot-prototype.appspot.com/OpenFOAM.html)

Explosion Shock Waves and High Strain Rate Phenomena
Materials Research Proceedings 13 (2019) 7-12

Materials Research Forum LLC
https://doi.org/10.21741/9781644900338-2

Numerical Simulation on Conical Shaped Charge with Copper Liner in Several Typical Shapes

Zhiyue Liu[1,a*], Junzhao Zhai[1,b] and Shuai Su[1,c]

[1]State Key Laboratory for Explosion and Technology, Beijing Institute of Technology, Zhongguancun South Street 5, Beijing 100081, P. R. China

[a]zyliu@bit.edu.cn, [b]toshiakichia@qq.com, [c]sushuai960930@gmail.com

Keywords: Shaped Charge, Copper Liner, Jet Formation, Penetration, Numerical Simulation

Abstract. Conical shaped charges with five types of liner geometries: (1) cone, (2) round-tipped cone (3) hemisphere, (4) ellipsoid, and (5) trumpet, have numerically been simulated to investigate jet formation and target penetration capabilities via LS-DYNA Multi-Materials Arbitrary Lagrangian Eulerian (MMALE) technique. The purpose is to observe the influences of liner shapes on performance of shaped charge setup and, the final behavior of such shaped charge devices. The explosive filling and the liner material are kept to be identical in the study. Simulation results show that the hemispherical liner brings out the lowest penetration performance, while the cone and round-tipped cone shaped liners result in highest performance, and, ellipsoid and trumpet liners are of middle performance, in comparison. Especially, shaped charges with both ellipsoid and trumpet liners present no remarkable discrepancy on penetration depth, but the entrance craters are of dramatically geometrical difference. Such penetration features are anticipated to be applicable in technical design of shaped charges for various specific applications.

Introduction

Jet from shaped charge device is of important role in penetration, cutting, perforation, and other applications [1]. Device of shaped charge is composed of two main parts: explosive filling and metallic liner. Traditionally, the conical shaped charge with conical copper liner has been utilized broadly. In the development of shaped charge skill, people pursued other geometrically shaped liners. Fedorov et al. [2] once proposed a liner with shape of hemisphere-cylinder combination for a shaped charge setup. Cao et al. [3] made numerical simulation on several shaped charges with different liner shapes and gained the conclusion that liner with hemisphere shape has worst penetration ability and hence limited application based on the simulated results of lowest jet tip velocity. In this paper, instead of experimental investigations, numerical technique is employed to study the performance of shaped charges with five kinds of geometrically shaped liners on the penetration abilities. The liners are of (1) cone, (2) round-tipped cone, (3) hemisphere, (4) ellipsoid, and (5) trumpet shapes, respectively. Multi-Materials Arbitrary Lagrangian Eulerian (MMALE) numerical technique in LS-DYNA software [4] is used to perform all numerical simulations. The focuses are made on the penetration depths on the target block and the sizes of the craters left by jet entrance. Such features may provide the useful support for the design of charges with suitable liner shape for specific application.

Simulation Procedures

Computational models. Shaped charge with cone liner is a commonest set-up in use. It is firstly selected for consideration. As depicted in Fig.1, it basically consists of a liner, an explosive filling, and an outer case [5]. The dimensions and materials relevant to the device are given in Table 1. A steel target, 15.24 cm diameater and 62.87 cm length, is set off the bottom of shaped

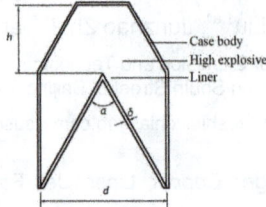

Fig. 1 Basic configuration of conical shaped charge considered for numerical simulation.

Table 1 Basic geometrical parameters of shaped charge model.

Liner thickness δ(cm)	Height of expl. filling h(cm)	Diameter d(cm)	Apex angle $\alpha(°)$
0.20	5.29	10.00	60

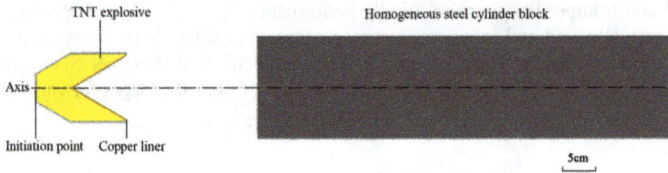

(a) Cone liner case

(b) Round-tipped cone liner case

(c) Hemisphere liner case

(d) Ellipsoid liner case

(e) Trumpet liner case

Fig.2 Assembly of shaped charge and witness steel block (a), and shaped charge devices possessing respective liner shapes (b-e).

Explosion Shock Waves and High Strain Rate Phenomena   Materials Research Forum LLC
Materials Research Proceedings **13** (2019) 7-12   https://doi.org/10.21741/9781644900338-2

charge with 20cm stand-off distance. The computational models for all five cases are illustrated in in Fig. 2.

Computational tools. Multiple Materials Arbitrary Lagrangian Eulerian (MMALE) solver in LS-DYNA software is used to for the whole computations.The characteristic size of mesh is set to be 0.05cm, and such size is used for all models. Quadrilateral meshes are utilized.

Equations of state and material models. For the high explosive, Jones-Wilkins-Lee (JWL) form equation of state is employed. The property parameters for TNT is found from Ref. [6]. For copper and aluminum, Steinberg strength model [7] is used for strength criterion, and Gruneisen equation of state is used under high pressure. The relevant parameters for copper and steel is obtained from Ref. [8]. Target is treated as dual-linear elastic-plastic material containing the criterion for failure. The relevant parameters used is from Ref. [5].

Results and discussion

Fig. 3 presents the computational results for the cone liner case, including jet formation, jet impinging onto target, as well as initial, middle, final penetrations in target block. Once the charge initiated, at 15μs, liner begins to collapse forming an initial jet shape. Utill the time of 62μs, the jet touches the target block as shown in Fig. 3(b). Fig. 3(c) displays the initial penetration of jets. With the later progress of penetration, jet velocity drops down and, jet mass is consumed by dispersion on the surface of the penetrated hole. At later phase, jet becomes slow, losing penetration ability just as Fig. 3(e) shows.

Similarly, simulation has also been performed for the cases with the other four liner shapes on jet formation and penetration. The detailed computational results are not further illustrated, only are typical features compared for those models. Firstly, jet tip velocities before striking the target are chosen for comparison and shown in Fig. 4. Among the five cases, shaped charge with round-tipped cone liner may provide the maximum jet velocity, following by the case with cone

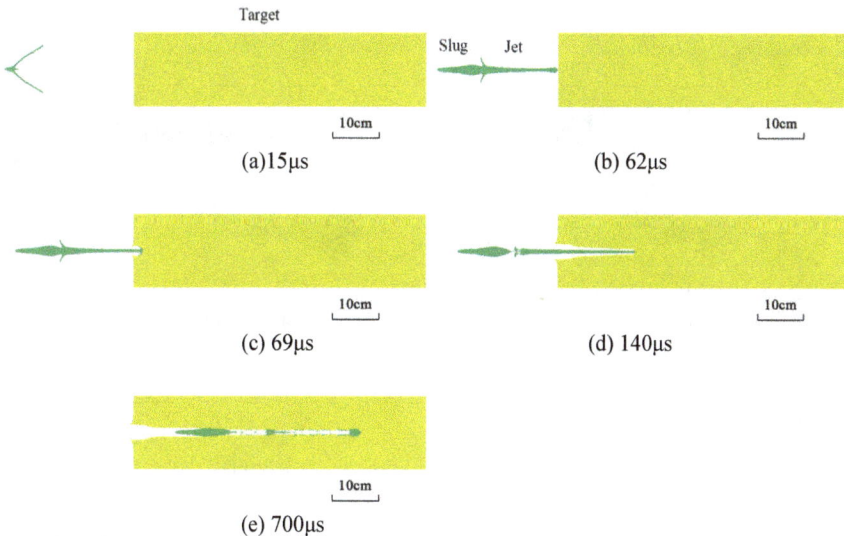

(a)15μs   (b) 62μs

(c) 69μs   (d) 140μs

(e) 700μs

Fig. 3 Jet formation and penetration into target by shaped charge with cone liner by simulation.

shape liner. Relatively, trumpet shape liner is also able to provide a higher jet velocity. However, ellipsoid shape liner brings out a slower jet velocity, and hemisphere shape liner produces the lowest jet velocity, only being higher than the half value of jet velocity from the round-tipped shape liner. In addition, with respect to the arrival time for the jet striking the target, trumpet shape liner is of the shortest one, then, followed by round-tipped cone liner; however, ellipsoid and

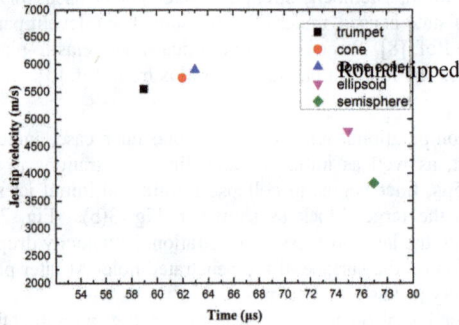

Fig. 4 Jet tip velocities just before striking the target by charges with various liners.

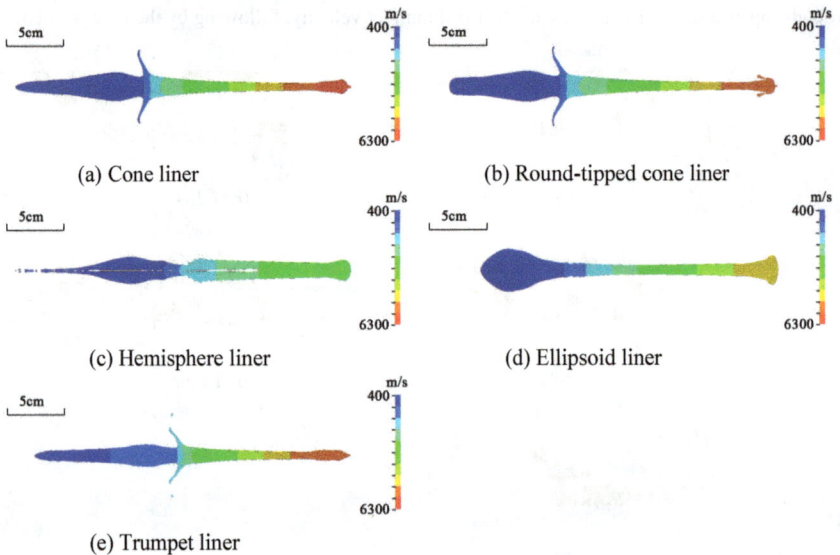

(a) Cone liner

(b) Round-tipped cone liner

(c) Hemisphere liner

(d) Ellipsoid liner

(e) Trumpet liner

Fig. 5 Configurations of jet and its velocity variation before striking the target from five charge models.

Explosion Shock Waves and High Strain Rate Phenomena   Materials Research Forum LLC
Materials Research Proceedings 13 (2019) 7-12   https://doi.org/10.21741/9781644900338-2

hemisphere liners exhibit much slower arrival time. The lag on the arrival time is due to a longer time for the collapse of liner to form jet for the cases of the ellipsoid and hemisphere liners.

To the penetration process, jet tip velocity is not the sole parameter to determine the depth of penetration. The important factor is kinematic energy of jet which completely determines the penetration depth. Kinematic energy of jet is related to mass distribution of jet as well as its velocity gradient. Fig. 5 shows the computed jet configurations for five charges with each liner shape in each charge before striking the target. The color variations in the figure denote the velocity magnitudes from 0.4km/s by blue to 6.3km/s by red. It clearly exhibits the velocity gradients in jet and slug during jet motion. The existence of velocity gradient would stretch the jet and force it to become slimmer. From Fig. 5 it is intuitively illustrated the jet tip velocities in all charge models. Charges with cone, round-tipped cone and trumpet liners produce a relatively higher tip velocity, and the cases with hemisphere and ellipsoid liners are of lower jet velocity distribution. However, in two charge models with hemisphere and ellipsoid liners, the jet diameter is extremely large, implying more mass distribution in jet portion.

Table 2 Calculated results of penetration depth by charges with different liner shapes.

Liner shape	Cone	Round-tipped cone	Hemisphere	Ellipsoid	Trumpet
Penetration depth (cm)	49.66	50.11	18.11	32.21	32.11

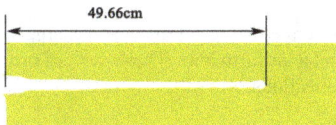

(a)Cone liner case

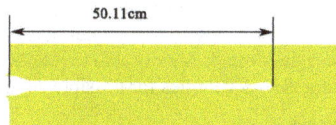

(b) Round-tipped cone liner case

(c) Hemisphere liner case

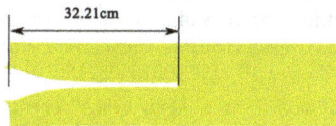

(d) Ellipsoid liner case

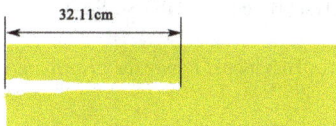

(e) Trumpet liner case

Fig. 6 Final penetration depths by jets from shaped charges with respective shape liners.

The total penetration depths are tabulated in Table 2 for all five charge models. Charge with hemisphere liner generates the lowest penetration depth of 18.11cm, however, both charges with cone and round-tipped cone liners even achieve the depth of around 50cm. Charges with ellipsoid and trumpet liners can produce the very similar penetration depth of 32.21cm and 32.11cm, respectively.

Fig. 6 shows the distributions of penetration holes by jets from five shaped charges as well as the graphical presentation of holes depths. Charges with cone and round-tipped cone liners produce holes with similar geometrical appearance and further, the penetration depths are very close. Charge with ellipsoid liner generates the very similar penetration depth, but the entrance craters are remarkably different. It implies that if large crater at entrance of jet is desired in the engineering application, charge with ellipsoid liner is a recommendable device. Moreover, charge with hemisphere liner produce the shortest penetration depth, however, the hole's diameter is large in total. It means if a larger and a uniform hole is required, this device is an ideal choice.

Conclusions

Jet formation and characteristics of target penetration from conical shaped charges with varied liner shapes have been simulated numerically via LS-DYNA dynamic software. The following conclusions can be achieved.

(1) Under the same mass of explosive charge, the traditional charge design with cone shape liner is able to produce the ideal penetration depth for targets. Other varied designs do not provide better penetration capability.

(2) Charge with ellipsoid shape liner may cause a larger crater at the entrance, and charge with hemisphere shape liner is able to produce a large and a uniform hole. Those designs provide enlarged opportunities for possible engineering applications.

References

[1] W. Walters and J. Zukas, Fundamentals of Shaped Charges, John Wiley & Sons, New York, 1989.

[2] S. V. Fedorov, Ya. M. Bayanova, S. V. Ladov, Numerical analysis of the effect of the geometric parameters of a combined shaped-charge liner on the mass and velocity of explosively formed compact elements, *Combustion, Explosion, and Shock Waves* 50(2015) 130-142. https://doi.org/10.1134/s0010508215010141

[3] L. Cao, X. Han, X. Dong, and Y. Zhang, Numerical Simulation of Effect of Liner Structure on Performance of Shaped Charge Jet, *MINING R & D* 29(2009) 98-100.

[4] LS-DYNA Manual R.9448, Livermore Software Technology Co., 2018.

[5] D. Shi, Y. Li, and S. Zhang, Explicit Dynamic Analysis Based on ANSYS/LS-DYNA, Tsinghua University Press, Beijing, 2005 (in Chinese).

[6] B. M. Dobratz and P. C. Crawford, LLNL Explosives Handbook, Lawrence Livermore National Laboratory Report, 1985.

[7] D. J. Steinberg, S. G. Cochran, and M. W. Guinan, A Constitutive Model for Metals Applicable at High Strain Rate, *J. App. Phys.* 51 (1980) 1498-1504. https://doi.org/10.1063/1.327799

[8] M. Katayama, S.Kibe, and T.Yamamoto, Numerical and Experimental Study on the Shaped Charge for Space Debris Assessment. *Acta Astronautica* 48 (2001) 363-372. https://doi.org/10.1016/s0094-5765(01)00073-x

Explosion Shock Waves and High Strain Rate Phenomena   Materials Research Forum LLC
Materials Research Proceedings **13** (2019) 13-24   https://doi.org/10.21741/9781644900338-3

Advanced Manufacturing under Impact / Shock Loading: Principles and Industrial Sustainable Applications

Academician Prof. Dr.-Ing. Dr.h.c. Prof.h.c. Athanasios G. Mamalis

Project Center for Nanotechnology and Advanced Engineering (PC-NAE),
NCSR "Demokritos", Athens, Greece

a.mamalis@inn.demokritos.gr

Keywords: Advanced Manufacturing, Advanced Materials, Net-Shape Manufacturing, Biomedical Engineering, Transport, Energy, Environment, Defense, Safety, Industrial Sustainability

Abstract. Trends and developments in *advanced manufacturing of advanced materials from macro- to nanoscale* subjected to static, lowspeed / high speed / hypervelocity impact and shock loading, with sustainable industrial applications to net-shape manufacturing, bioengineering, transport, energy and environment, defense and safety, an outcome of the very extensive, over 50 years, work on these scientific and industrial areas performed by the author and his research international team, are briefly outlined. The impact of such advanced materials, manufacturing and loading techniques, products and applications on many technological areas, e.g. the manufacturing/machine tool sector, communications / data storage, transportations, health treatment, energy conservation, environmental and human-life protection, is significant and highly beneficial.

Introduction

The topics considered, an outcome of the very extensive academic and industrial work over 50 years on these fields performed by the author and his research international team, may be listed as:

- Mechanics (Structural plasticity, Low / High speed impact loading, Hypervelocity impact, Shockwaves loading)
- Precision / Ultraprecision manufacturing from macro-, micro- to nanoscale (Metal forming, Metal removal processing, Surface engineering / Wear, Non-conventional techniques)
- Nanotechnology / Nanomaterials manufacturing
- Ferrous and non-ferrous materials (Metals, Ceramics, Superhard, Polymers, Composites, Multifunctional), from macro- to nanoscale (Nanostructured materials, Nanoparticles, Nanocomposites)
- Powder production and processing technologies (High strain-rate phenomena and treatment under shock: Explosives, Electromagnetics, High temperature / high pressure techniques)
- Biomechanics / Biomedical engineering
- Transport / Crashworthiness of Vehicles: Passive and active safety for passengers and cargo (Surface transport: Automotive, Railway; Aeronautics: Aircraft, Helicopters)
- Energy (Superconductors, Semiconductors, Electromagnetics, Solar cells, Photovoltaics, Nuclear reactors)
- Environmental aspects (Impact on climate change: Nanotechnology; Automotive industry; Aeronautics industry)
- Safety (Detection of explosives and hazardous materials)
- Defense (Ballistics, Projectiles hitting targets, Shock loading)

Explosion Shock Waves and High Strain Rate Phenomena   Materials Research Forum LLC
Materials Research Proceedings **13** (2019) 13-24   https://doi.org/10.21741/9781644900338-3

- Industrial sustainability

Some trends and developments in *Advanced Manufacturing from macro- to nanoscale* in the important engineering topics from industrial, research and academic point of view: nanotechnology, precision /ultraprecision engineering and advanced materials (metals, ceramics, polymeric, composites/nanocomposites) under static, low/high speed impact, hypervelocity impactand shock loading, with *sustainable industrial applications* to net-shape manufacturing, bioengineering, transport, energy/environment and defense / safety, are briefly outlined in the present ESHP 2019 Invited Lecture.

Manufacturing Technology Principles

The *principles* of *advanced manufacturing technology* may be identified by six main elements, see Fig. 1, with the central one being the enforced *deformation* to the material, i.e. the *processing* itself, brought about under consideration of the *interface between tool and workpiece*, introducing interdisciplinary features for lubrication and friction, tool materials properties and the surface integrity of the component. The *as-received material structure* is seriously altered through the deformation processing, subjected from *static* to *very high-strain rate phenomena / shock loading*, therefore, materials testing and quality control *before and after processing* are predominantly areas of interest to the mechanics, manufacturing and materials scientists. The performance of the *machine tools* together with the *tool design* are also very important, whilst, nowadays, the *techno-economical aspects*, like the notion of *manufacturing systems,* e.g. automation, modeling and simulation, rapid prototyping, process planning, computer integrated manufacturing, *energy conservation* and *recycling*, as well as *environmental aspects* are important in *advanced manufacturing engineering* [1].

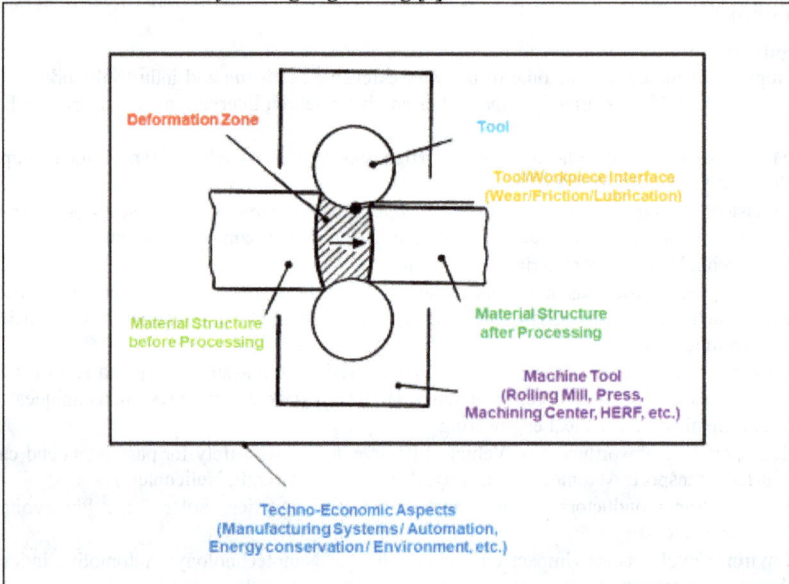

Fig. 1 Advanced manufacturing technology principle

Explosion Shock Waves and High Strain Rate Phenomena
Materials Research Proceedings 13 (2019) 13-24

Materials Research Forum LLC
https://doi.org/10.21741/9781644900338-3

The *structural plasticity mechanics*, governing the *deformation* of the material, see Fig. 2, are mainly associated with [1, 2]:

(a) *Low strain-rate phenomena*, i.e. deformation under *static-, low speed impact loading*, for metals, polymers and composite materials, see Fig. 2(i).

In this case, the material behavior is characterized by its stress-strain curve. Ductile metals and polymers are plastically deformed with the formation of stationary and traveling plastic hinges.

Contrary to this ductile mechanism, the deformation mechanism for brittle composite materials is achieved by material fragmentation developing extensive microcracking processes easily controlled and depended on the properties of fibers and resins the fibers orientation.

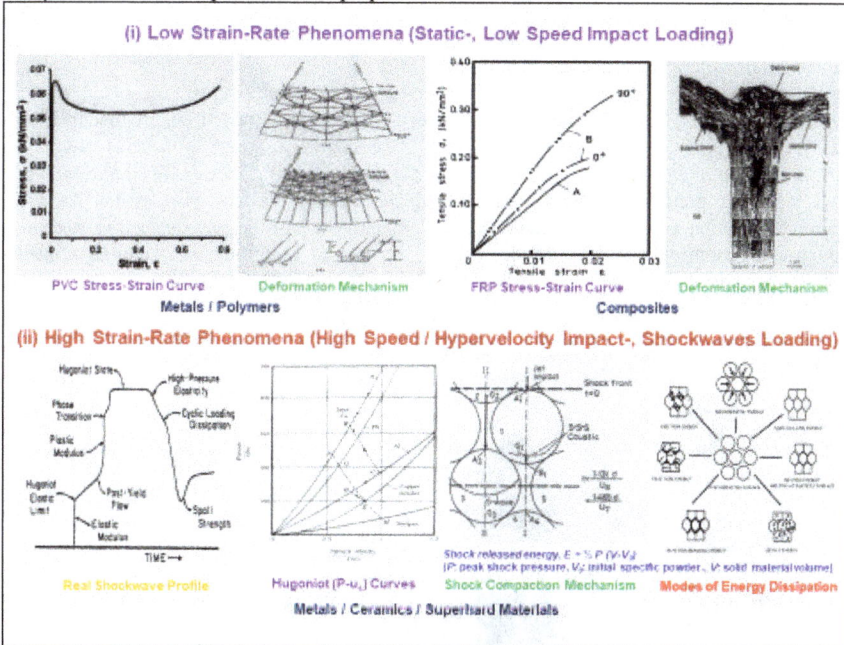

Fig. 2 *Structural plasticity mechanics*

(b) *High strain-rate phenomena*, i.e. deformation under *high speed / hypervelocity impact-, shockwaves loading)*, for metals, ceramics and superhard materials (diamonds, CBN), see Fig. 2(ii).

During *dynamic / shock loading*, a *longitudinal, P-shockwave*, with a real shockwave profile (pressure, P vs time, t), is initiated, traveling into the body at high speed, calculated from the corresponding state of the material under shock conditions, i.e. its Hugoniot curve (pressure, P - specific volume, V relationship), defined as the loci of all shock states and essentially describing the material properties. The particles are accelerated into the pores at high velocities, impacting each other, which results in the development of *shear S-waves* in the particles due to jet impact at a point on the particle surface, traveling inside the particle and reflected at its surface resulting in jet formation due to spalling, with subsequent loading of the already formed jet moving between the interparticle voids in the same direction as the shock. The frictional energy release results, therefore, in melting at the surface regions with the associated bonding once the material is solidified. In the consolidation of brittle materials, particle fracture also occurs, leading to the filling of the gaps, whilst reactive elements can also be added to help the bonding process. The high-pressure state creates numerous lattice defects and dislocation substructures leading very often to localise shearing and microcracking. The energy dissipation modes due to shockwaves and the relevant mechanisms, are related to the *shock released energy*, $E = \frac{1}{2} P (V-V_0)$, where P is the peak shock pressure, V_0 the initial specific powder volume and V the volume of the solid material.

Quality of manufactured parts is mainly determined by their dimensional and shape accuracy, the surface integrity, and the functional properties of the products. Development of manufacture engineering is related to the *tendency to miniaturization* and is accompanied by the *continuous increasing of the accuracy of the manufactured parts*. The two main trends towards the miniaturization of products are, see Fig. 3:

• *Precision/ Ultraprecision manufacturing* (Metal forming, Metal removal processing, Surface engineering / Wear, Non-conventional techniques), see Fig. 3(i), carried out by machine tools with very high accuracy;

• *Nanotechnology processing*, see Fig. 3(i), i.e. the fabrication of devices with atomic and / or molecular scale precision by employing *new advanced energy beam processes* that allow for atom manipulation and therefore, the design and manufacture of the *nanostructured materials*, having every atom or molecule in a designated location and exhibiting novel and significantly improved physical, chemical, mechanical and electrical properties.

The various stages of *nanomaterials manufacturing* are listed in Figure 3(ii) [3].

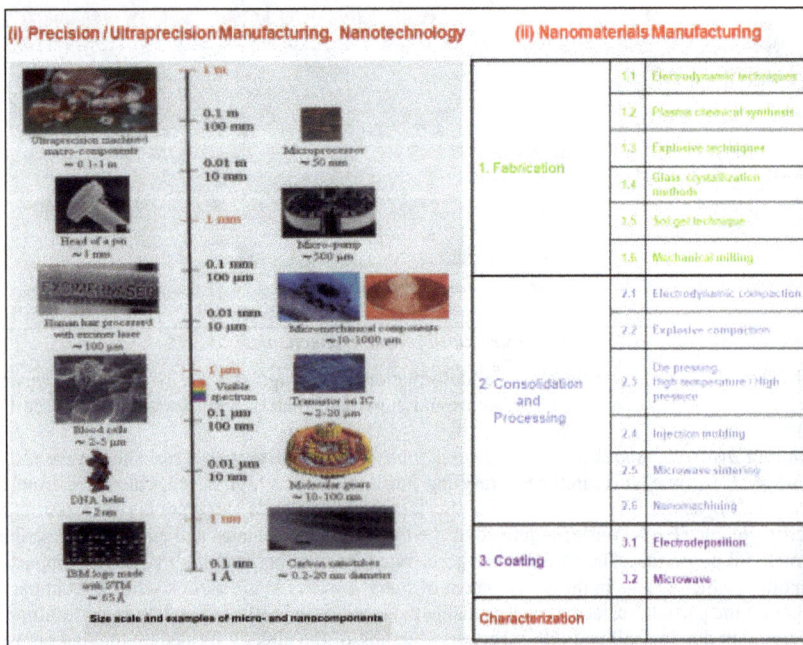

Fig. 3

Explosion Shock Waves and High Strain Rate Phenomena   Materials Research Forum LLC
Materials Research Proceedings **13** (2019) 13-24   https://doi.org/10.21741/9781644900338-3

Macro-, micro- and nanoproducts under shock loading involve the production of ultrafine materials (metals, ceramics, mixtures) by explosive and/or electromagnetic compaction as well as high temperature /high pressure die pressing techniques, see Fig. 4 [4]. These materials can be utilized as a basic material or as the modifying additives at the manufacturing of sintered powdery ceramics, hard-alloy and ceramic composites, nanoparticles reinforced metal and polymeric, matrix composites, abrasive pastes and suspensions and polishing ones, chemical catalysts and sorbents.

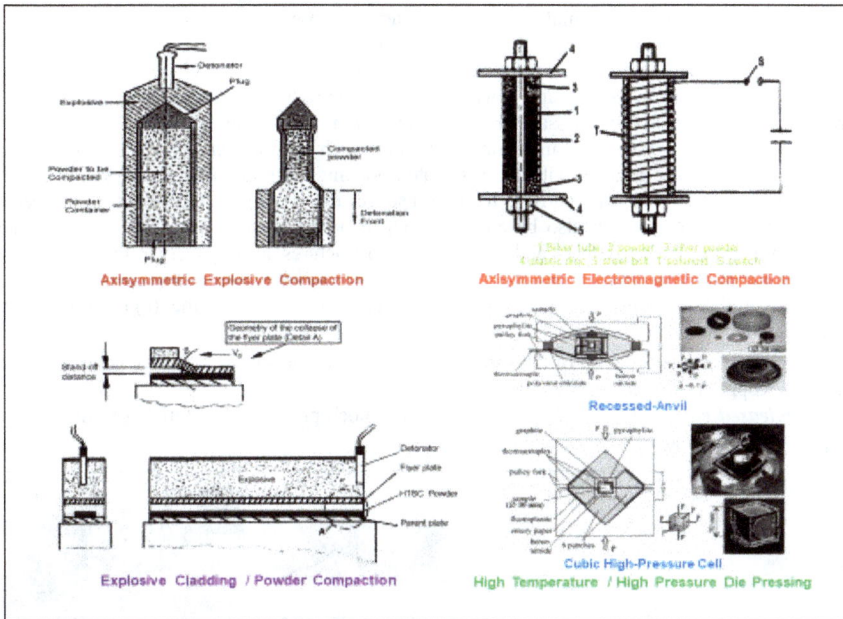

Fig. 4 Net-shape manufacturing under shock loading

Industrial Sustainable Applications
Sustainability nowadays is challenged on many fronts, among the most prominent ones, the advanced manufacturing may be considered. Industrial sustainable applications in advanced manufacturing from macro- to nanoscale of advanced materials (metals, ceramics, superhard materials, polymers, composites / nanocomposites) under static, low- / high-speed / hypervelocity impact and shock loading, an outcome of the very extensive, over 50 years, academic and industrial work on these fields of the author in association with his vast research international team worldwide, are briefly listed below.

(a) Powder production and processing technologies
   (High strain-rate phenomena and treatment under shock: Explosives, Electromagnetics, High temperature / high pressure techniques).
   Crushing of brittle materials grains can be achieved by *shock loading*. These high-strain rate phenomena are used for producing materials of micro- and nanoscale grains. By applying

properly calculated and directed shock waves created by explosion, Al_2O_3, MgO, ZrO_2, Mo, Ti, metal MgB_2 and ceramic high-Tc superconductors are treated for reducing their grain size into nanoscale.

Compaction of such materials by shockwaves has the advantage that during the compaction phase grain growth does not occur. The high-speed shockwaves with high energy content can be created either by *initiating high explosives* (*explosive compaction*) or by *discharging electric capacitors* (*electromagnetic compaction*). In successfully consolidated products, interparticle regions that are molten and rapidly solidified are usually observed, which is more profound in *metals* than in *ceramics*. The main defect of compacted ceramics is the *presence of cracks*, propagating through the whole component, that can be eliminated by novel compaction designs, powder preheating or even by using reactive mixtures to produce heat by exothermic reaction, triggered by the shock wave passage. During *compaction*, the powder surfaces are accelerated into the pores at high velocities, impacting each other, with frictional energy release, leading to melting at the surface regions with the associated bonding once this material is solidified. In consolidated brittle materials, particle fracture also occurs, leading to the filling of the gaps, whilst reactive elements can also be added to help bonding process. The high-pressure state creates numerous lattice defects and dislocation substructures leading very often to localise shearing and microcracking.

(i) Consolidation mechanism stages and particle shape changes after time: (a) t=0 (beginning of impact); (b) t=1.465d/U_T; (c) t=2·1.465d/U_T+ t_c, see Fig. 5(i).

(ii) Experimental validation of the shock compaction mechanism by explosive compaction of spherical copper powders, see Fig. 5(ii).

Shock released energy, $E = \frac{1}{2} P (V-V_0)$ (*P*: peak shock pressure; V_0: initial specific powder volume; *V*: volume of solid material).

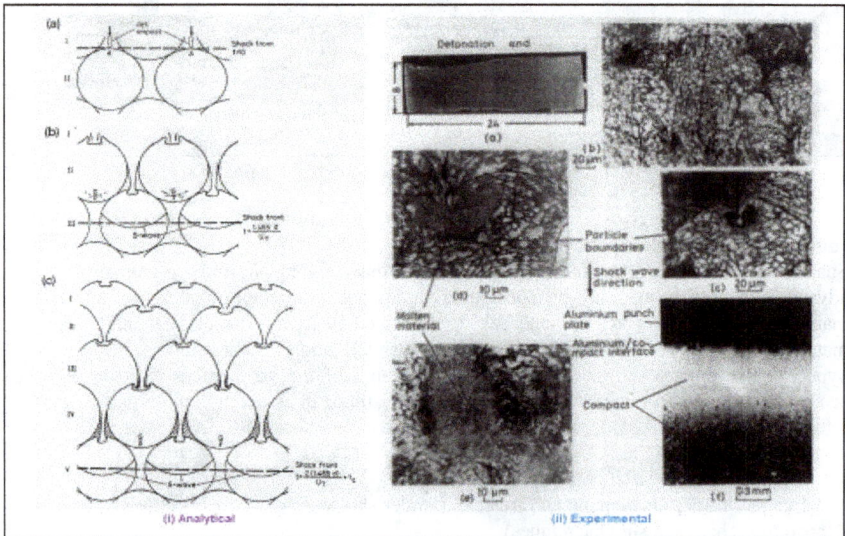

Fig. 5 Consolidation mechanism

(b) *Biomechanics / Biomedical engineering*

Two industrial sustainable cases are considered, see Fig. 6:

(i) *Shock production* of *nanodiamonds doped with boron* is obtained by detonating highexplosives, at detonation velocities up to 7 km/s, in an explosive chamber. Industrial applications of these nanodiamonds are related to*ultraprecision nanoprocessing / nanolithography*, with nanodiamonds used as multifunctional Scanning Tunneling Microscope Berkovich pyramidshaped tips, and to *biomedical engineering*, with diamond nanoplatforms employed for targeted delivery of diagnostic and therapeutic agents in *oncology*, see Fig. 6(i) [5].

(ii) *Sapphire head / sapphire cup / titanium stem hip-joint endoprostheses*, produced by ultraprecision manufacturing, constitute an important *biomechanics* sustainable application in *biomedical orthopedics* and are associated with the *active safety* for the passengers protection during crash, see Fig. 6(ii) [6].

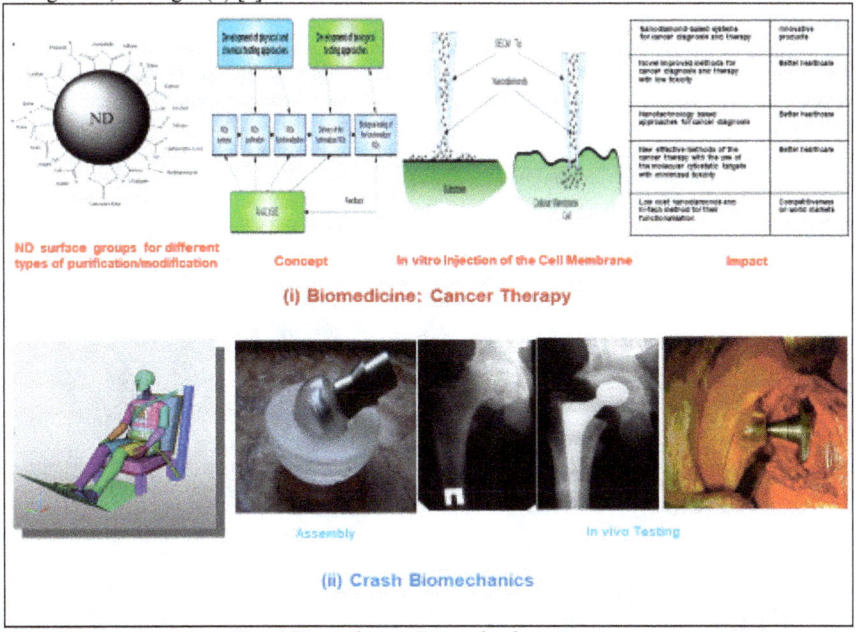

Fig. 6 Biomechanics/Biomedical engineering

(c) *Energy*

Topics considered are: Super- and Semiconductors, Electromagnetics, Solar Cells / Photovoltaics and Nuclear Reactors. In particular, materials that present zero resistance at a certain critical temperature, T_c above the absolute zero are named Superconductors and the related phenomenon Superconductivity. High-temperature superconducting materials of the YBCO and BSCCO ceramic compounds (T_c=77-93K), fabricated by various physicochemical techniques (solid state reaction, sol-gel, etc.) in the form of powders / nanoparticles, and Low-temperature superconducting nanostructured MgB_2 metallic materials (T_c=49K), whilst high-energy rate powder compaction processes, e.g. explosive and electromagnetic compaction and high pressure / high temperature techniques, and subsequent forming and metal removal processing, are used to manufacture superconductive ceramics and metals with unique properties, with applications to electricity and transport, see Fig. 7(i) [7].

(d) *Environmental aspects: Impact on climate change*

Nanotechnology as well as the *automotive* and the *aeronautics* industries are among the factors that have a profound impact on energy consumption and hence greenhouse gas emissions (GHG).

- *Aeronautics Industry Impact on Climate Change*: A novel design of *5th generation gas turbine*

engines for *transport* and *energy*, using *structural ceramics blades* manufactured by *self-propagating high-temperature synthesis* and *electrochemical dimensional processing*, is proposed, reducing the effect of thermo-mechanical and dynamic loading.

- The *technological impact* of nanotechnology on climate change as well various *strategies* to Combat it, see Fig. 7(ii), by following the *energy supply chain*, i.e. *renewable energies*; *photovoltaics* and *solar energy, wind energy, fuel cells* and *hydrogen economy; energystorage, batteries, supercapacitors*; *thermoelectric conversion efficiency* in combustion and electric engines; *weight reduction* by using lighter, stronger and stiffer nanocomposite materials with the potential to significantly reduce dead weight and promote energy efficiency in transportation, and, furthermore, by utilising nanotechnology applications involving nanomanufacturing, use of bulk and surface nanomaterials and nanomaterials of biological origin or interacting with living organisms, may be helpful inside a reduction global climate change mitigation effort [8].

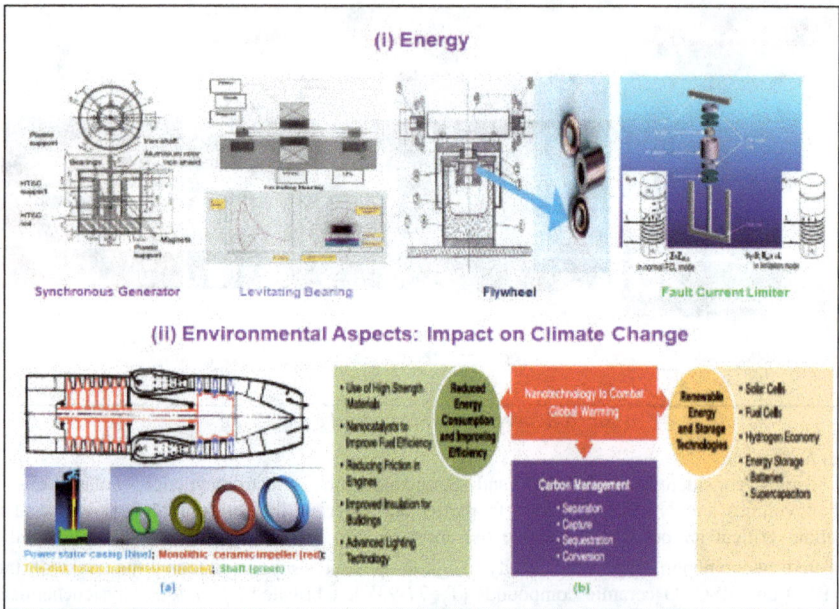

Fig. 7

(e) *Transport / Crashworthiness of Vehicles: Passive and active safety for passengers / cargo*

Crashworthiness studies provide with the mechanism by which a proportion of impact energy is absorbed by the collapsing structure, whilst a small amount is transferred to the passenger in order to improve the crash resistance of the vehicle. To obtain effective crashworthy behavior, associated with the *passive safety*, a Crashworthiness study must be carried out in the very early design stages, considering analytical and numerical modeling and experimental in situ and at laboratory scale of thin-wall structural components subjected to various loading conditions, in particular, *low and high speed impact*. Note, also, that, *activesafety* is associated with the passengers and cargoprotection during crash. For passengers, it is mainly related to *biomedical engineering*, with the *hip-joint endoprostheses* being an important *crash biomechanics* application, see Fig. 6(ii). My, over 40 years, extensive scientific and industrial work on the *crash mechanism* of metals, polymers, composite materials and advanced hybrid composite structures, related to *surface transport, automotive* and *railway*, and *aeronautics* (*aircraft, helicopters*), is briefly outlined in Figs. 8(i), 8(ii) and 8(iii), respectively [9].

The relation between CO_2 *emissions from road vehicles, trains, ships* and *aircrafts*, attributed to each sector of the human activity, and *the climate change* is established, *Electric* and *hybrid cars* considerably reduce emissions resulting in both global warming and air pollution locally, and, in addition, to help curb the world's dependence on oil. Sustainable technologies to reduce or totally eliminate the impact of vehicles on climate change are: *Hybrid Electric Vehicles (HEV), Plug-in Hybrid Electric Vehicles (PHEV), Battery Electric Vehicle with Range Extender (REBEV), Battery Electric Vehicles (BEV)* and *Fuel Cell Electric Vehicles (FCEV)*, seeFig. 8(i).

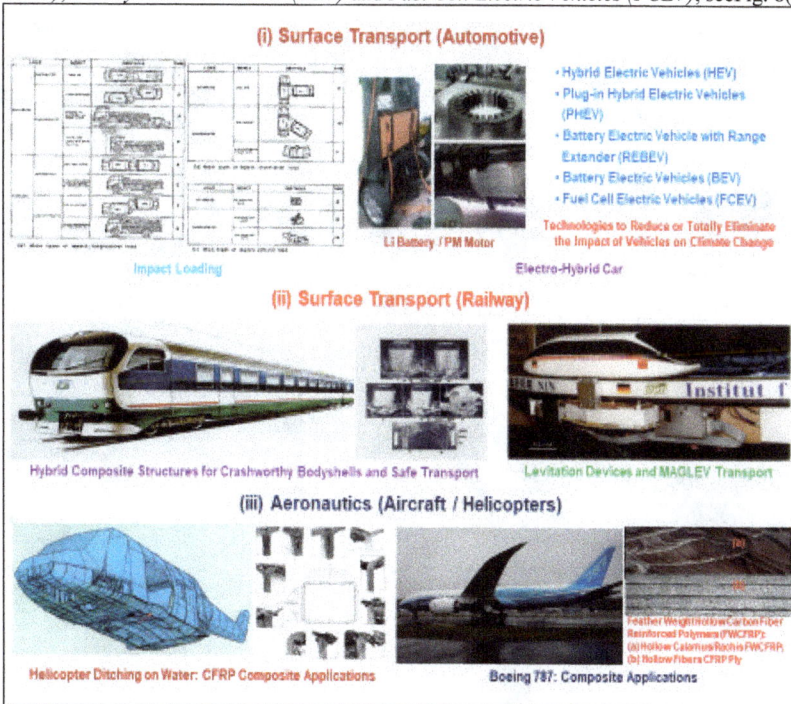

Fig. 8

(f) *Defense*

High strain-rate phenomena related to *hypervelocity impact loading* are employed in the case of *ballistics* and *projectiles / bullets hitting targets* for the proper design and manufacture of the *armory* in Military installations, e.g. *tanks, weapons* etc., see Fig. 9(i) [2].

(g) *Safety*

Prevention of terrorist attacks is of utmost importance, with *explosives* being the chosen weapons targeting any populated area. Three processes are necessary for the detection of explosives and hazardous materials in the air: *collectivity*, consisting of front-end collection and pre-concentration; *separation* and *detection,* providing *selectivity* and *sensitivity* of the threat, respectively. It results to the manufacture of a *multi-channel explosive detection sensor* with different sets of sorbents that provide separation of various groups of explosives, consisting of carbon nanotubes in combination with various monomers-organic compounds effectively interacting with certain nitro-aromatics and using a diamond plate as substrate for the measuring matrix due to the chemical inertness and stable physical-mechanical characteristics of the diamond.

Development of new methods against *improvised explosive devices (IEDs)* and *home-made explosives (HMEs)* and improvement of existing facilities for faster, more sensitive and cheaper explosive detection and neutralization of person- or vehicle-borne IEDs, are the objectives of the *Integrated Network for the Detection of Explosives(INDEX)* concept, see Fig. (ii) [10].

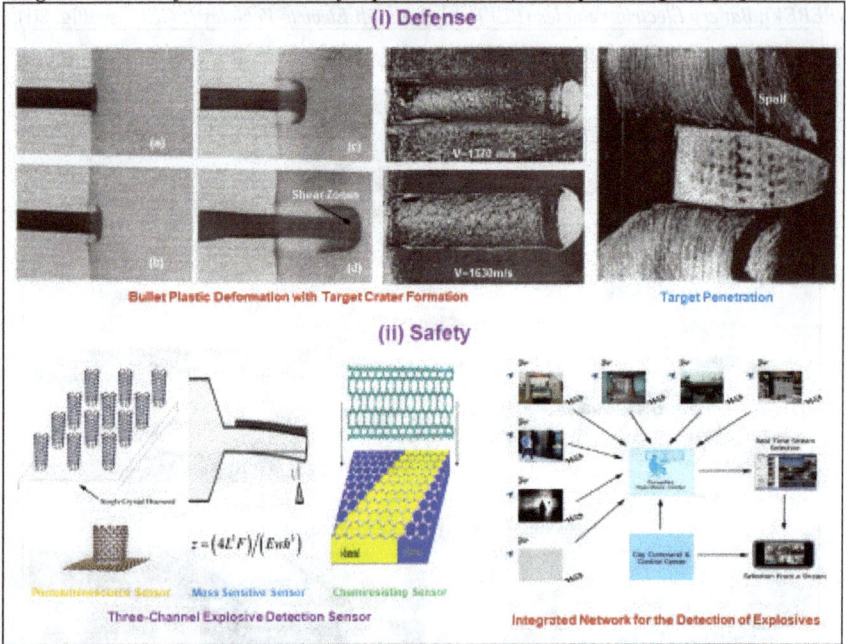

Fig. 9

Explosion Shock Waves and High Strain Rate Phenomena
Materials Research Proceedings 13 (2019) 13-24

Materials Research Forum LLC
https://doi.org/10.21741/9781644900338-3

Concluding Remarks

The benefits of such *advanced materials*, *manufacturing* and *loading techniques*, *products* and *industrial sustainable applications* in many technological areas are significant. The impact of these technologies in every day's life is considered to be great, since it will make the *manufacturing / machine tool sector, communications, transportations, data storage, health treatment, energy conservation, environmental* and *human-life protection* and many other technological applications *better, faster, safer, cleaner* and *cheaper.*

At this point, it is necessary to clearly identify the value of a reasonably sound knowledge of the kind presented so far:

Industry is conducted both for profit-making and at the same time for supplying goods to the mass of the people at minimum, economic prices; primary concern is the manufacturing cost.

Defence industries or requirements are concerned with the product to fulfil a certain task and cost tends to be secondary.

To help carry out these two functions efficiently, *Research programs* have usually to be put in operation and, in order to do this at minimum cost, the long, costly ladder of research and development must be mounted at as high a level as possible. This is most easily done by enhancing *international research cooperation* between *Research Centers, Universities* and *Industry*. In that respect, the *PC-NAE*, continuing the over 40 years established international cooperation worldwide of my *Laboratory of Manufacturing Technology of the NTUA*, is encouraging it by expanding all these cooperation activities with the establishment of *Multinational Clusters*, like the rather newly established *Shockwaves Cluster*. Quoting the Great Ancient Greek Philosopher Aristotle:

ΑΡΙΣΤΟΤΕΛΗΣ *(384-323 BC)*
«Αναλυτικά Ύστερα»

**"Knowledge of the fact is different from
knowledge of the reason for the fact"**

References

[1] Johnson, W., Mamalis A.G., *Engineering Plasticity: Theory of Metal FormingProcesses*, Springer Verlag (CISM Courses and Lectures No 139), Wien, 1977, pp. 345

[2] Johnson W, MamalisA.G., *Gegenüberstellung statischer und dynamischer Schadens oder Deformationserscheinungen*, Fortschritt-Berichte der VDI-Zeitschriften, Reihe 5, Nr.32, Düsseldorf, 1977, pp. 78

[3] Mamalis A.G., *Recent advances in nanotechnology*, Journal of Materials Processing, Technology, 2007. **181**: pp52-58

[4] Mamalis A.G., *Powder processing*. International Journal of Production Engineering and Computers, 2003. **5**(6): pp 15-31

[5] Lysenko O.G., Grushko V.I., Dub S.N., Mitshevich E.I., Novikov N.V., MamalisA.G., *Manufacturing and characterization of nanostructures using Scanning Tunneling Microscopy with diamond tip.* Journal of Nano Research, 2016. **42**: pp. 14-46, https://doi.org/10.4028/www.scientific.net/jnanor.42.14

[6] Mamalis A.G., Lytvynov K.A., Filipenko V.A., Lavrynenko S.N., Ramsden J.J., Soukakos P.N., *Perfection of contemporary hip-joint endoprostheses by using a sapphire-sapphire friction pair.* Journal of Biological Physics and Chemistry, 2007. **7**: pp. 3-5. https://doi.org/10.4024/10701.jbpc.07.01

[7] Mamalis A.G., Szalay A., Manolakos D.E., Pantazopoulos G., *Processing of High Temperature Superconductors at High Strain Rates,*Technomic Publishing Co,USA. 2000, pp. 276, https://doi.org/10.1201/9781420014266

[8] Mamalis A.G., Ramsden J.J., Holt G.C., Vortselas A.K., Mamali A.A., *The effect of nanotechnology in mitigation and adaptation strategies in response to climate change.* Nanotechnology Perceptions, 2011. **7**: pp. 159-179, https://doi.org/10.4024/n08mal1a.ntp.07.03

[9] Mamalis A.G., Manolakos D.E., Demosthenous G.A. and Ioannidis M.B., *Crashworthiness of Composite Thin-Walled Structural Components*, Technomic Publishing Co, USA, 1998, pp. 269

[10] Lysenko O.G., Grushko V.I., Mitshevich E.I. and Mamalis A.G., *Three channel trace explosive detector using diamond.* (to be published)

Explosion Shock Waves and High Strain Rate Phenomena   Materials Research Forum LLC
Materials Research Proceedings **13** (2019) 25-30   https://doi.org/10.21741/9781644900338-4

High Strain Rate Behaviour of Auxetic Cellular Structures

Nejc Novak[1, a *], Matej Vesenjak[1, b], Shigeru Tanaka[2, c], Kazuyuki Hokamoto[2, d], Baoqiao Guo[3, e], Pengwan Chen[3, f] and Zoran Ren[1,4 g]

[1] Faculty of Mechanical Engineering, University of Maribor, Maribor, Slovenia

[2] Institute of Pulsed Power Science, Kumamoto University, Kumamoto, Japan

[3] State Key Laboratory of Explosion Science and Technology, Beijing Institute of Technology, Beijing, China

[4] IROAST, Kumamoto University, Kumamoto, Japan

[a]n.novak@um.si, [b]matej.vesenjak@um.si, [c]tanaka@mech.kumamoto-u.ac.jp, [d]hokamoto@mech.kumamoto-u.ac.jp, [e]baoqiao_guo@bit.edu.cn, [f]pwchen@bit.edu.cn, [g]zoran.ren@um.si

Keywords: Cellular Structures, Auxetic Materials, High Strain Rate, Experimental Testing, Computational Simulations

Abstract. Auxetic cellular structures are modern metamaterials with negative Poisson's ratio. The auxetic cellular structures build from inverted tetrapods were fabricated and experimentally tested under dynamic loading conditions to evaluate the effect of strain rate on their deformation mode. The Split-Hopkinson Pressure Bar (SHPB) apparatus was used for testing at strain rates up to 1,250 s^{-1}, while a powder gun was used for testing at strain rates up to 5,000 s^{-1}. The homogeneous deformation mode was observed at lower strain rates, while shock deformation mode was predominant at higher rates. The results have shown that the strain rate hardening of analysed auxetic specimens is prominent at higher strain rates when the shock deformation mode is observed, i.e. when most of deformation occurs at the impact front. Relevant computational models in LS-DYNA were developed and validated. A very good correlation between the computational and experimental data was observed.

Introduction

Auxetic cellular structures are novel metamaterials with negative Poisson's ratio – they tend to expand in lateral direction when subjected to tensile loading and vice versa in the case of compression loading [1]. This behaviour can be beneficial in many applications, especially in the crashworthiness, ballistic protection and energy absorption applications [2]. The mechanical behaviour of auxetic structures is well characterised and understood for quasi-static loading conditions, but not so much for dynamic and impact loading due to insufficient experimental characterisation attempts so far. Past studies were mostly concerned with the quasi-static elastic behaviour of uniform auxetic structures at very small strains [3] and limited ballistic resistance study [4]. Mechanical behaviour of some particular auxetic structures was characterised by uniaxial quasi-static compressive and tensile tests [5–10]. The Split-Hopkinson Pressure Bar (SHPB) experiments were also carried out for auxetic cellular structures fabricated with additive manufacturing, including also polymer fillers [11]. There is a clear need to test the auxetic cellular structures also under high strain rate loading conditions to comprehensively evaluate their behaviour also at highly dynamic loading. Especially since there are many applications where these metamaterials can be used efficiently.

Explosion Shock Waves and High Strain Rate Phenomena   Materials Research Forum LLC
Materials Research Proceedings **13** (2019) 25-30   https://doi.org/10.21741/9781644900338-4

Specimens fabrication

The specimens build from inverted tetrapods were used in this research. Inverted tetrapods (Fig. 1a), are assembled in a particular way to define the geometry of the investigated specimens (Fig. 1b-c). The specimen's inverted tetrapod dimensions were (Fig. 1a): $a = 3.5$ mm, $h = 3$ mm, $dh = 0.5$ mm, while the circular cross-section diameter of the struts was in range from 0.38 to 0.53, depending on the porosity (Table 1). Two types of specimens were analysed in this work: a) short and b) long specimens. The difference between the analysed types of specimens was in length in X2 direction (Table 1). The specimens were fabricated from the Ti-6Al-4V alloy powder by the selective electron-beam melting method (SEBM) at the Institute of Materials Science and Technology (WTM), University of Erlangen-Nürnberg, Germany [12].

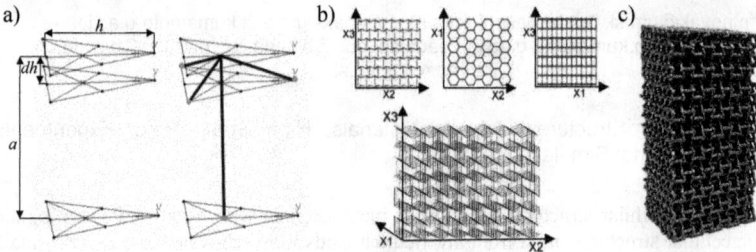

Figure 1: Geometry of auxetic specimens build from inverted tetrapods: a) inverted tetrapod, b) geometry in orthogonal views and c) fabricated specimen

Table 1: Specimens data

	Short specimens			
	Dimensions [mm] $\lvert X1\rvert \times \lvert X2\rvert \times \lvert X3\rvert$	Mass (std. dev.) [g]	Density [g/cm^3]	Porosity p [-]
Middle porosity	$15.6 \times 19.2 \times 18.7$	3.45 (0.023)	0.62	0.86
	Long specimens			
	Dimensions [mm] $\lvert X1\rvert \times \lvert X2\rvert \times \lvert X3\rvert$	Mass (std. dev.) [g]	Density [g/cm^3]	Porosity p [-]
High porosity		4.67 (0.138)	0.40	0.91
Middle porosity	$15.6 \times 40.5 \times 18.7$	7.24 (0.051)	0.61	0.86
Low porosity		9.04 (0.075)	0.76	0.83

High strain rate experimental testing

High strain rate experimental testing was performed using two different experimental devices: a) Split Hopkinson Pressure Bar (SHPB) apparatus at the State Key Laboratory of Explosion Science and Technology, Beijing Institute of Technology, Beijing, China and b) the powder gun at the Institute of Pulsed Power Science, Kumamoto University, Kumamoto, Japan. The achieved loading velocities were 25 m/s and 220 m/s using the SHPB and the powder gun, respectively. The loading velocities correspond to the strain rates up to 1,250 s^{-1} and 10,000 s^{-1} for short specimens and 5,000 s^{-1} for long specimens, respectively. In the case of powder gun experiments the mechanical response was evaluated with the PVDF gauge (Piezo film stress gauge, PVF2 11-125EK, Dynasen), as in previous experiments described by Tanaka et al. [13]. A homogenous

Materials Research Forum LLC
https://doi.org/10.21741/9781644900338-4

deformation mode of short auxetic specimens was observed (similar to quasi-static loading conditions) behaviour during the SHPB loading, Fig. 2 [10].

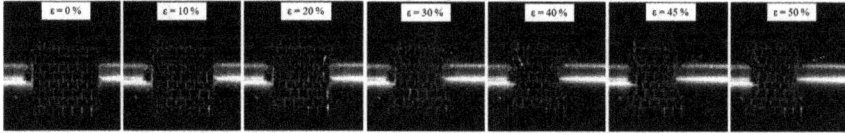

Figure 2: Deformation behaviour of short specimens in X1 direction during SHPB testing (view in the X2 direction)

The typical deformation behaviour of long specimens during the powder gun experiment is shown in Fig. 3, where the crushing sequence was analysed using digital images of the high-speed camera recording. Due to high strain rates achieved (10,000 s^{-1} and 5,000 s^{-1}), the shock deformation mode was observed in all tested specimens with various porosities. The main deformation of specimens throughout the tests (up to very large deformations) was concentrated only at the area, where the specimens impacts the fixed rigid plate.

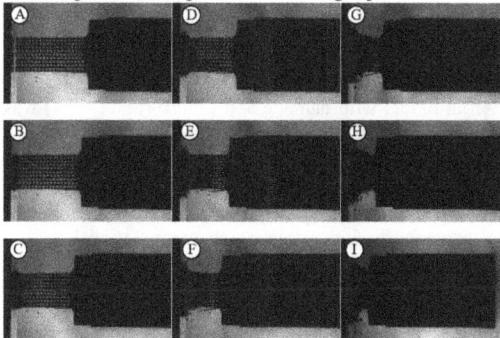

Figure 3: Deformation behaviour of long specimens with middle porosity in X2 direction during powder gun testing

The average pressure (stress) measurements recorded by the gauge at the fixed rigid measuring plate for long specimens with all three porosities and their comparison to quasi-static results are shown in Fig. 4. A signal filter with moving point average (10 data points) was used to smooth higher frequency response in recorded experimental data. As it can be observed from Fig. 4, the change of deformation mode (from homogenous mode under quasi-static loading to shock mode under high strain rate loading) causes large strain rate hardening in comparison to the results of quasi-static tests (with a strain rate of 0.005 s^{-1}) for identical auxetic specimens [10].

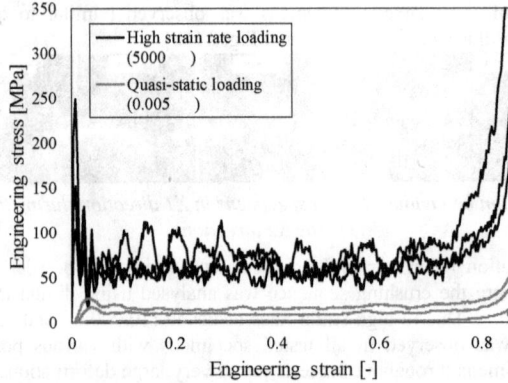

Figure 4: Comparison of quasi-static and high strain rate loading responses of long auxetic cellular structures with different porosities

Computational simulations

The relevant computational models were developed using the LS-DYNA software and validated. The elasto-plastic material model with damage and failure (MAT_024) was validated using the above reported experimental results. The relevant material parameters for the applied material model are the following: density ρ = 4430 kg/m^3, Young's modulus E = 120000 MPa, Poisson's ratio v = 0.3, initial yield stress σ_{yield} = 1300 MPa, the definition of linear hardening with the second point in stress-strain diagram (σ_2, $\varepsilon_{pl,2}$) – (1800 MPa, 0.8) and the critical strain ε_{crit} = 0.15 at the start of material damage. Further details about computational modelling can be found in previously reported work [10, 14].

The experimental results obtained by SHPB were used only for visual comparison of deformation behaviour. An excellent agreement between the experimental and computational results can be observed from Fig. 5.

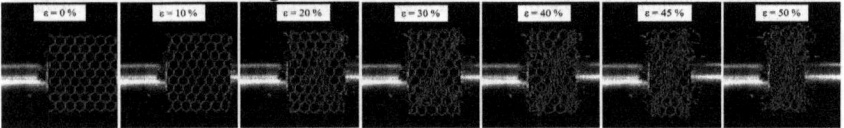

Figure 5: Comparison of experimental and computational results of SHPB loading of short specimens in X1 direction (view in the direction X3)

The visual comparison of computational and experimental results for the case of powder gun loading is shown in Fig. 6, where good agreement can be observed as well. The computational model predicts the change of deformation mode from homogeneous to shock mode successfully.

The results from the computational models were further compared with experimental results in terms of engineering stress (Fig. 7), where a good correlation was noted. The mechanical properties (e.g. densification strain and plateau stress) are reproducible in the case of low and middle porosity, while in the case of high porosity, a larger deviation can be observed. This phenomenon was already explained and justified in previous work [12], where the consequence of the fabrication procedure and resulting imperfections was discussed in detail.

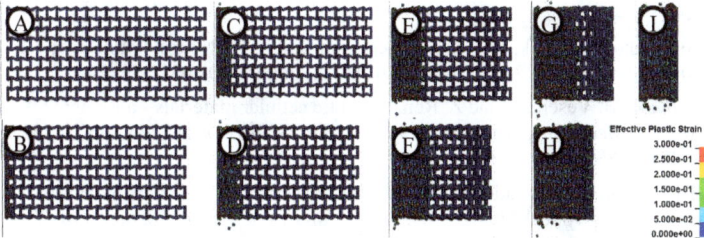

Figure 6: Deformation behaviour of long specimens with middle porosity in X2 direction in powder gun testing

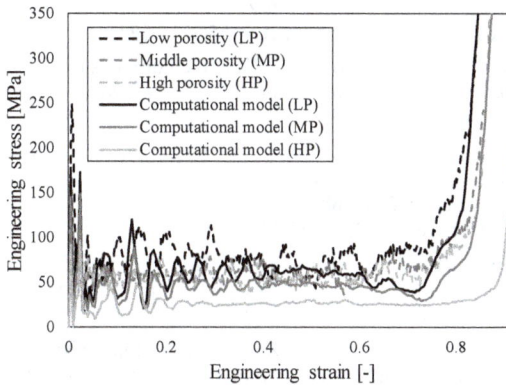

Figure 7: Comparison of experimental and computational results of powder gun loading of long specimens with different porosities

Summary

The auxetic cellular structures build from inverted tetrapods were analysed in this work. The high strain rate deformation response was evaluated with experimental testing using the SHPB apparatus and the powder gun device. The homogeneous deformation mode was observed in the case of SHPB testing at strain rates up to up to 1,250 s^{-1}. In the case of powder gun loading at strain rates up to up to 10,000 s^{-1} the deformation mode changes to shock mode, where most of the material is deformed at the impact front. The change of deformation mode causes large strain rate hardening effect of all analysed auxetic structures, which significantly influences the mechanical response. The relevant computational models were developed and successfully validated.

Acknowledgement

The research was performed within the framework of the basic research project J2-8186 entitled "Development of multifunctional auxetic cellular structures", which is financed by the Slovenian Research Agency "ARRS". The authors acknowledge the financial support from the Slovenian Research Agency (research core funding No. P2-0063). This paper is supported by the opening project of State Key Laboratory of Explosion Science and Technology (Beijing Institute of Technology) - opening project number KFJJ17-03M.

References

[1]   K. E. Evans, M. A. Nkansah, I. J. Hutchinson, and S. C. Rogers, "Molecular Network Design," *Nature*, vol. 353, p. 124, 1991. https://doi.org/10.1038/353124a0

[2]   N. Novak, M. Vesenjak, and Z. Ren, "Auxetic cellular materials - a Review," *Strojniški Vestn. - J. Mech. Eng.*, vol. 62, no. 9, pp. 485–493, 2016. https://doi.org/10.5545/sv-jme.2016.3656

[3]   T. Bückmann *et al.*, "Tailored 3D Mechanical Metamaterials Made by Dip-in Direct-Laser-Writing Optical Lithography," *Adv. Mater.*, vol. 24, no. 20, pp. 2710–2714, 2012. https://doi.org/10.1002/adma.201200584

[4]   C. Qi, S. Yang, D. Wang, and L.-J. Yang, "Ballistic resistance of honeycomb sandwich panels under in-plane high-velocity impact," *Sci. World J.*, vol. 2013, no. SEPTEMBER 2013, p. 892781, 2013. https://doi.org/10.1155/2013/892781

[5]   E. A. Friis, R. S. Lakes, and J. B. Park, "Negative Poisson's ratio polymeric and metallic foams," *J. Mater. Sci.*, vol. 23, no. 12, pp. 4406–4414, 1988. https://doi.org/10.1007/bf00551939

[6]   S. Mohsenizadeh, R. Alipour, M. Shokri Rad, A. Farokhi Nejad, and Z. Ahmad, "Crashworthiness assessment of auxetic foam-filled tube under quasi-static axial loading," *Mater. Des.*, vol. 88, pp. 258–268, Dec. 2015. https://doi.org/10.1016/j.matdes.2015.08.152

[7]   S. Hou, T. Liu, Z. Zhang, X. Han, and Q. Li, "How does negative Poisson's ratio of foam filler affect crashworthiness?," *Mater. Des.*, vol. 82, pp. 247–259, Oct. 2015. https://doi.org/10.1016/j.matdes.2015.05.050

[8]   L. Yang, O. Harrysson, H. West, and D. Cormier, "Compressive properties of Ti–6Al–4V auxetic mesh structures made by electron beam melting," *Acta Mater.*, vol. 60, no. 8, pp. 3370–3379, 2012. https://doi.org/10.1016/j.actamat.2012.03.015

[9]   L. Yang, O. Harrysson, H. West, and D. Cormier, "Mechanical properties of 3D re-entrant honeycomb auxetic structures realized via additive manufacturing," *Int. J. Solids Struct.*, vol. 69–70, pp. 475–490, Sep. 2015. https://doi.org/10.1016/j.ijsolstr.2015.05.005

[10]   N. Novak, M. Vesenjak, L. Krstulović-Opara, and Z. Ren, "Mechanical characterisation of auxetic cellular structures built from inverted tetrapods," *Compos. Struct.*, vol. 196, pp. 96–107, 2018. https://doi.org/10.1016/j.compstruct.2018.05.024

[11]   T. Fíla *et al.*, "Impact Testing of Polymer-filled Auxetics Using Split Hopkinson Pressure Bar," *Adv. Eng. Mater.*, p. n/a--n/a, 2017. https://doi.org/10.1002/adem.201700076

[12]   J. Schwerdtfeger, P. Heinl, R. F. Singer, and C. Körner, "Auxetic cellular structures through selective electron beam melting," *Phys. Status Solidi B*, vol. 247, no. 2, pp. 269–272, 2010. https://doi.org/10.1002/pssb.200945513

[13]   S. Tanaka *et al.*, "High-velocity impact experiment of aluminum foam sample using powder gun," *Meas. J. Int. Meas. Confed.*, vol. 44, no. 10, pp. 2185–2189, 2011. https://doi.org/10.1016/j.measurement.2011.07.018

[14]   N. Novak, K. Hokamoto, M. Vesenjak, and Z. Ren, "Mechanical behaviour of auxetic cellular structures built from inverted tetrapods at high strain rates," *Int. J. Impact Eng.*, vol. 122, pp. 83–90, 2018. https://doi.org/10.1016/j.ijimpeng.2018.08.001

Explosion Shock Waves and High Strain Rate Phenomena
Materials Research Proceedings 13 (2019) 31-34

Materials Research Forum LLC
https://doi.org/10.21741/9781644900338-5

Shock Synthesis of Gd$_2$Zr$_2$O$_7$

Toshimori Sekine[1*], Qiang Zhou[2], Pengwan Chen[2], Zhen Tan[2], Haotian Ran[2], and Jianjun Liu[3]

[1]Center for High Pressure Science & Technology Advanced Research, P.R. China

[2]Beijing Institute of Technology, P.R. China

[3]Beijing University of Chemical Technology, P.R. China

* toshimori.sekine@hpstar.ac.cn

Keywords: Shock Synthesis, Pyrochlore, Fluorite, Weberite, Solid Reactions, Gd$_2$Zr$_2$O$_7$

Abstract. We dealt with shock compression on a composition of Gd$_2$Zr$_2$O$_7$ by explosive-driven flyer impact methods, because Gd$_2$Zr$_2$O$_7$ with r$_{Gd}$/r$_{Zr}$ ration of 1.46 lies at the structural boundary between ordered pyrochlore and defect fluorite structures. The results indicate recovered products depend on shock conditions that we need to specify by further study.

Introduction

Shock compression process provides unique environments for materials synthesis due to not only the realized high pressure and high temperature but also the shock-enhanced kinetics and fast quenching [1]. The process is a time-limited reaction and favors martensitic phase transformation in general. There are many trials to use shock compression techniques for investigation of solid state reactions [2]. The most typical one has been known historically as diamond synthesis, and the process has been developed to optimize the yield of products. Here we report a progress of shock synthesis of oxide compounds using explosive-driven plate impacts.

Rare earth pyrochlore compounds of A$_2$B$_2$O$_7$, where A is a rare earth element and B is a tetravalent cation such as Zr^{4+} and Ti^{4+}, exhibit several interesting properties for physical, chemical and industrial applications. The pyrochlore structure is known to form if the cation radii ratio (r$_A$/r$_B$) lies in the range 1.46–1.80. However, the fluorite structure is favored with r$_A$/r$_B$ below 1.46. The cation radii ratio r$_A$/r$_B$ has an important effect on the high pressure structural stability. Gd$_2$Zr$_2$O$_7$ with r$_A$/r$_B$ ratio of 1.46 lies at the structural boundary between ordered pyrochlore and defect fluorite. Hence it is expected to show interesting structural behavior as a function of temperature and pressure. We tried to understand the effect of shock compression on Gd$_2$Zr$_2$O$_7$. Among A$_2$B$_2$X$_7$ (X is anion such as O and F) compounds there are three discrete structures of pyroclore, fluorite, and weberite. Their structural relations are based on the fluorite structure (AX$_2$) where each anion is at the center of the cation tetrahedral (A$_4$X) and the lattice is characterized by a lattice constant of a = ~5 Å with Z=1. In pyrochlore structure, different A and B cations make A$_4$X, B$_4$X, and A$_2$B$_2$X, and the lattice is expanded double (a = ~10 Å) and the number of Z=8. Weberite consists of A$_3$BX, AB$_3$X, and A$_2$B$_2$X, with lattice constants of √2a, 2a, and √2a and with Z=4. Therefore, pyrochlore and weberite have their corresponding superlattices in addition to the fluorite structure.

Shock compression technique has never been applied to solid-solid reactions in complicate chemical systems to our knowledge. We explore such chemical systems using shock compression techniques.

Explosion Shock Waves and High Strain Rate Phenomena
Materials Research Proceedings **13** (2019) 31-34

Materials Research Forum LLC
https://doi.org/10.21741/9781644900338-5

Experimental methods

We dealt shock compression on two starting mixtures of a composition $Gd_2Zr_2O_7$ (powdered mixture of Gd_2O_3 + 2 ZrO_2 and the product heated in air at 900 °C for 2 hours), encapsulated in copper containers, by explosive-driven flyer impact methods [3]. A copper flyer with a diameter of 40 mm and a thickness of 2 mm is accelerated to a high velocity by the detonation of the main explosive charge of nitromethane (CH3NO2), initiated by a booster charge of 8701 explosive [3]. Peak shock pressure reflected within a sample is calculated by the impedance match method from the known impact velocity. The shock velocity (Us km/s)- particle velocity (Up km/s) relation of copper [4] with density of 8.924 g/cm^3) is used as Us = 3.91 + 1.51Up. The starting material preheated at 900 °C (sample I) was partially reacted to a fluorite structure with monoclinic ZrO_2 according to the powder x-ray diffraction (XRD) analysis (Fig. 1 A, B, and C). Another starting material (sample II) was a mixture of Gd_2O_3 and monoclinic ZrO_2 as the received powders.

The container after shot was cut open to remove the sample. The successfully recovered samples as well as the starting materials were investigated by powder x-ray diffraction methods to identify phases present in products. We carried out a series of recovery experiments as a function of impact velocity and porosity.

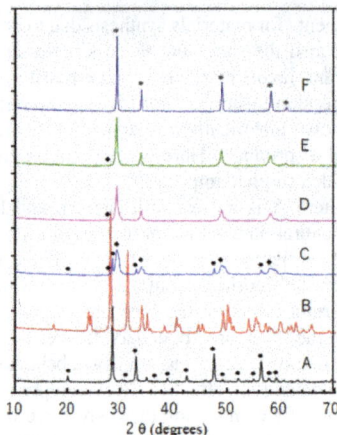

Fig. 1. XRD patterns using Cu *Kα* radiation
for sample I.
(A) Gd_2O_3 after heated at 1000°C for 4 hours,
(B) ZrO_2 after heated at 800°C for 4 hours,
(C) Product from a mixture of (A) and 2 (B) after
heated at 900°C for 2 hours,
(D) Recovered sample I with porosity of 48%
at 43.5 GPa, (E) Recovered sample I with porosity
of 40% at 60.0 GPa, and (F) Recovered sample I
with porosity of 54% at 84.9 GPa.

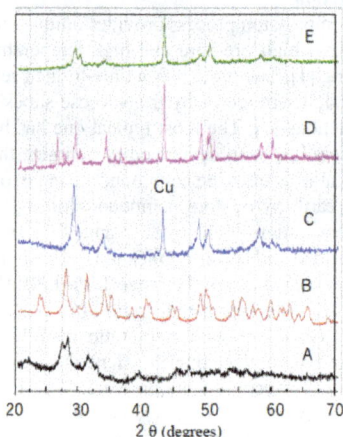

Fig. 2. XRD patterns using Cu *Kα* radiation
for sample II.
(A) Gd2O3 as received, (B) ZrO2 as received,
(C) Recovered sample II with porosity of 50%
at 60.0 GPa, (D) and (E) Recovered sample II
with porosity of 30% at 84.9 GPa. Peak with Cu
Indicates the highest peak for Cu.

Explosion Shock Waves and High Strain Rate Phenomena   Materials Research Forum LLC
Materials Research Proceedings 13 (2019) 31-34   https://doi.org/10.21741/9781644900338-5

Results and discussion

We have started a series of shock recovery experiments on various pyrocholre compounds that has not known yet, and explore novel compounds using shock compression. We report and discuss the results on two bulk compositions of $Gd_2Zr_2O_7$. Impact velocities of 1.85 km/s, 2/36 km/s, and 3.06 km/s of Cu flyers correspond to peak shock pressures of 43.5 GPa, 60.0 GPa, and 84.9 GPa, respectively.

The XRD patterns of recovered samples indicate completely transformed to the fluorite structure with no additional peaks (Fig. 1 F) from sample I. The peaks sharpen with increasing shock pressure (Figs. 1 D and E). The starting sample (Fig. 1 C) indicates peaks corresponding to a fluoride structure and Gd_2O_3 with no ZrO_2. Although this result suggests that the fluorite is non-stoichiometric, the initial monoclinic ZrO_2 may be transformed to cubic or tetragonal structure at shock-induced high temperatures. The maximum peaks for fluorite and tetragonal (or cubic) ZrO_2 are close around 30 degree each other, and the difference between tetragonal and cubic ZrO_2 is indistinguishable by XRD [5].

The products from the raw powder mixture (sample II), however, display two types of XRD patterns (Fig. 2). One consists of relatively broad peaks corresponding to a pyrochlore structure (Figs. 2 C and E) and the other indicate relatively sharp peaks of pyrochlore structure (Fig. D), although both contain significant amounts of copper powders from container and may contain small amount of tetragonal (or cubic) ZrO_2. The copper contamination that we did not observed in sample I may suggest higher shock temperatures in sample II than sample I because the starting Gd_2O_3 powder was poorly crystalline (Fig. 2 A).

Then, the presence of a large amount of copper in the recovered sample II can be explained by high temperatures, although the porosity difference may affect the shock temperature. The formation of fluorite $Gd_2Zr_2O_7$ suggests relatively high temperatures (>1530°C) in hot press sintering [6]. If this is the case, our shock temperatures could be close to this. The effects of porosity of the initial powders pressed in the recovery container are not well controlled in the present study, and we need further study. However, high temperatures generated in powdered samples are found to promote solid reaction significantly. It is difficult to understand the shock pressure effect on the solid reaction due to a small difference between fluorite and pyrochlore structures at high pressures. Based on a detailed study of the lattice parameter of $Gd_2Zr_2O_7$ with fluorite and pyrochlore structures at ambient condition [7], the pyrochlore has slightly larger volume than the fluorite and can be the low pressure. Therefore, we need to know shock conditions to understand the solid reactions. And also it is interesting to compare the static compression results on $Gd_2Zr_2O_7$ at room temperature [8, 9]. The results indicate back transformation from pyrochlore to defect-fluorite formed above 15 GPa [8] and amorphization above ~35 GPa due to distortion of cation [9].

Summary

Shock compressions of powders with a composition of $Gd_2Zr_2O_7$ produced both defect fluorite and ordered pyrochlore structures detected by x-ray diffraction methods. The results need to be specified to understand the solid reactions.

References

[1] T. Sekine, Experimental methods of shock wave research for solids. *In* Hypervelocity Launchers (ISBN: 978-3-319-26016-7) *Shock Wave Science and Technology Reference Library* Vol. 10, F. Seifler and O. Igra (Eds), 55-76, Springer. 2016. https://doi.org/10.1007/978-3-319-26018-1_3

[2] S.S. Batsanov, Features of solid-phase transformations induced by shock compression. Rus. Chem. Rev. 75 (2006) 601-616. https://doi.org/10.1070/rc2006v075n07abeh003613

[3] Z. Tan , P. Chen, Q. Zhou, J. Liu, X. Mei, B. Wang, and N. Cui, Shock synthesis and characterization of titanium dioxide with α-PbO_2 structure J. Phys.: Condens. Matter 30 (2018) 264006. https://doi.org/10.1088/1361-648x/aac709

[4] S.P. Marsh, LASL Shock Hugoniot Data, University of California Press (1980).

[5] R. Srinivasa, R.J. De Angelis, G. Ice, and B.H. Davis, Identification of tetragonal and cubic structures of zirconia using synchrotron x-radiation source. J. Mat. Res., 6 (1991) 1287-1292. https://doi.org/10.1557/jmr.1991.1287

[6] U. Brykala, R. Diduszko, K. Jach, and J. Jagielski, Hot pressing of gadolinium zirconate pyrochlore. Ceramic Intern., 41(2015) 2015-2021. https://doi.org/10.1016/j.ceramint.2014.09.114

[7] Y.H. Leea, H.S. Sheub, J.P. Denga, and H.-C.I. Kao, Preparation and fluorite– pyrochlore phase transformation in Gd2Zr2O7. J. Alloy. Comp., 487 (2009) 595-598. https://doi.org/10.1016/j.jallcom.2009.08.021

[8] F.X. Zhang, J. Lian, U. Becker, R.C. Ewing, J. Hu, and S.K. Saxena, High-pressure structure changes in the $Gd_2Zr_2O_7$ pyrochlore. Phys. Rev. B 76 (2007)214104.

[9] N.R. Sanjay Kumar, N.V. Chandra Shekar, and P.Ch. Sahu, Pressure induced structural transformation of pyrochlore $Gd_2Zr_2O_7$. Solid St. Comm., 147 (2008) 357-359. https://doi.org/10.1016/j.ssc.2008.06.028

Explosion Shock Waves and High Strain Rate Phenomena   Materials Research Forum LLC
Materials Research Proceedings 13 (2019) 35-40   https://doi.org/10.21741/9781644900338-6

Experimental Study for the Tenderness of Meat using Underwater Shock Waves Generation by Wire Electrical Discharges

Ken Shimojima [1, a *], Yoshikazu Higa [1,b], Osamu Higa [1,c], Ayumi Takemoto [1,d], ,
Hideaki Kawai [2,e], Kazuyuki Hokamoto [3,f] Hirofumi Iyama[4,g],
Toshiaki Watanabe [5,h] and Shigeru Itoh[6,i]

[1]Nat. Inst. Tech.Okinawa College of Technology,905 Henoko, Nago Okinawa905-2192, Japan

[2]Asahi Giken, Tokura1-17-19 Toyonaka Osaka 561-0845, Japan

[3]Kumamoto Univ.,2-39-1 Kurokami Chuo-ku Kumamoto 860-8555,Japan

[4]Nat. Inst. Tech., Kumamoto College, Shisuya2659-2 Koushi Kumamoto 861-1102, Japan

[5] National Fisheries Univ., 2-7-1 Nagata-Honmachi, Shimonoseki 759-6595, Japan, Japan

[6] Inst. Shockwave Applied Technology,2070-19 Otohime Aso Kumamoto 869-2226,Japan

[a] k_shimo@okinawa-ct.ac.jp, [b] y.higa@okinawa-ct.ac.jp, [c]osamu@okinawa-ct.ac.jp,
[d]tkmt@okinawa-ct.ac.jp, [e] h.kawai@asahigiken-kk.com, [f] hokamoto@mech.kumamoto-u.ac.jp,
[g] eyama@kumamoto-nct.ac.jp, [h] watanabe@fish-u.ac.jp, [i]itoh_lab@okinawa-ct.ac.jp

Keywords: Underwater Shock Wave, Meat Softening

Abstract. High age of the population advances in the world. The consumption of meat increases. Some methods of softening of edible meat are methods such as electric energy, pressure, heating and biological. The development of the method of the tenderness that is the high efficiency which can apply to the volume of production of the meat is expected. The National Institute of Technology, Okinawa College (OkNCT) has developed a food processing machine that generates underwater shock waves through wire electrical discharge. The machine can be used for sterilization, milling, tenderness, and extraction among others. In this study, we experimentally examined the conditions for food tenderness using pork as the food material in the experiments. The relationship of the tenderness of edible meat measured with a durometer with the number of underwater shock wave generation, and the distance of the shock wave generation point from the edible meat and reflectance backing material were shown experimentally. The prototype design of the pressure vessel for the processing of the meat was shown. The possibility of designing and manufacturing of a pressure vessel according to the required tenderness was shown.

1. Introduction

In USA, Europe, Japan, the consumptions of edible meat increase [1-3]. The population of the elderly has been increasing in Japan. The Japanese increase rate is serious, but almost increases in all developed countries [4]. There have been many studies on the tenderness of food [5-6]. Tenderness is one of the most important sensory characteristics of meat. The conventional method of tenderness meat include pounding food materials or cutting muscle fibers with a knife or Biological method etc. [7]. It is shown that the softening of the meat using hydrodynamic shock wave. However, hydrodynamic shock wave is produced by an explosive and is unsuitable for practical use [8]. Biochemical techniques have shown the possibility of tenderness of meat [9-13].It is showed that particularly the application of the proteolytic enzymes transglutaminase and phrases, associated with nutritional, technological, and environmental improvements[14].

I.N.A. Ashie et al. studied on the softening of meat using an aspartic proteinase (AP) and papain on meat proteins. An improvement of 20 to 30% was confirmed in the softening of the meat [15]. 'SLOW cooking' was shown to be effective for soft meat dishes [16]. Highly cold-shortened muscle indicated the possibility of tenderness of a meat [17,18]. However, it is necessary to develop mechanical devices for practical use by these new food processing method. The possibility of changing the softening of meat was shown by using high concentration of oxygen. However, processing has several days. A processing method in which the processing time is short and the design and production of a mechanical apparatus that can for practical use is realistic is effective. At OkNCT, a food processing machine that generates underwater shock waves has been developed. This device consists of a high-voltage power-supply unit for shock wave generation and a pressure vessel for processing. The effects of this processing are improvement of extractability, tenderness, and sterilization without heating [19-24]. We develop the device for processing meat using shock waves. However, the conditions of processing using shock waves and their relationships with the tenderness of meat have not been clarified. Clarification of the relationships is necessary for the design of the pressure vessel most suitable for meat processing. In this study, the relationships of the number of underwater shock wave processing and the distance of an underwater shock wave generation point from meat with the tenderness of meat were examined experimentally. The clarification of these relationships may also lead to the understanding of this mechanism of the tenderness of meat. A guideline for the design of the pressure vessel for processing meat is provided. Figure 1 shows an overview of the food processing device that generates underwater shock waves.

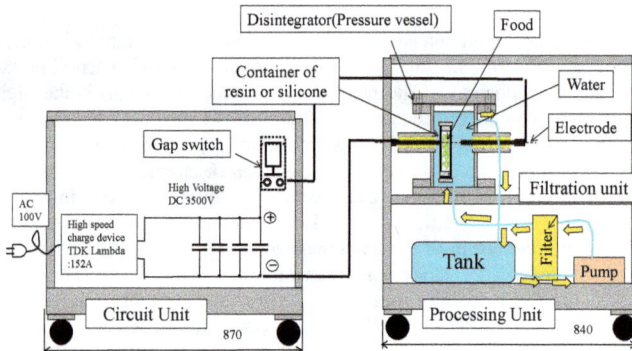

Fig.1 Figure of food processing machine for test processing

Electric energy is charged with the hi-charging device (TDK Lambda: 152A) from a switchboard (200V, 20A).The pressure vessel for food processing is filled with water. The charged electric energy is supplied to an electrode in the pressure vessel by the gap switch with the air cylinder. A thin aluminum wire is installed between electrodes, and it is evaporated by thermite reaction induced by instantaneously applied high voltage, resulting in shock wave generation. The generated shock wave propagates in the water in the pressure vessel. The food to be softened is packed with silicone or resin. The food is destroyed at the interface owing to the density difference resulting in spalling-destruction-phenomenon. The meat is processed (softened) by this phenomenon.

Explosion Shock Waves and High Strain Rate Phenomena   Materials Research Forum LLC
Materials Research Proceedings **13** (2019) 35-40   https://doi.org/10.21741/9781644900338-6

2. Experimental conditions

2.1 Processing conditions for tenderness using the underwater shock waves

The pressure of underwater shock waves is exponentially decreases with transmission distance[25].To clarify the relationship between meat tenderness and the distance of the shock wave propagation point from the meat, meat tenderness was measured at various distances between the meat and the shock wave generation point. The effects of these waves may be unhanced by increasing the number of times shock waves are generated using this device experimentally. To clarify the relationship between the number of processing and tenderness, the tenderness of meat was measured after one to three underwater shock waves with the distance of the shockwave generation point from the meat (135mm). We chose some materials of the backing material. To clarify the relationship between the backing material used and meat tenderness, meat tenderness was measured using different backing materials at various distances of the shockwave generation point from the meat (135mm). Figure 2 shows a picture of the pressure vessel for experiments on tenderness.

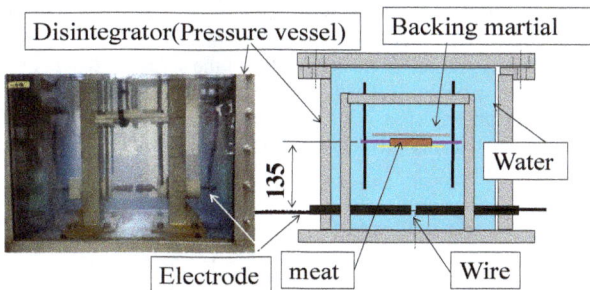

Fig.2 Picture of pressure vessel for experiment

The size of the vessel is 1m cubic. This meat placed underwater is supported by a silicone sheet. The distance between the shock wave generation point and the meat is changed by screws. We prepared water (reflectance to water,0%), wood (83.6%), and stainless steel (92.6%) as backing materials. The other parameters were constant. The charge voltage was 3.5KV, the condenser capacity was 800μF (This is connected by four parallel). The materials of thin line is aluminum. The diameter of the thin line was 1mm, the distance of the thin line at each electrodes was 8mm.

2.2 Method of measurement of tenderness

We chose pork, as the material for studying the tenderness effect on edible meat. Because tenderness differs depends on part of meat, we processed meat after cutting the red meat part. Figure 3 shows the measurement points on the meat. The red meat of pork shown in the figure was 10cm in width, 12cm in height, and 1cm in thickness. The measurement points were indicated by red food dye. The measurement points were located as follows: one point at the center of the meat, one each above and below the center point , two points each on the right and left sides of the center point. The distance between points wad 1cm. The tenderness of each point was measured after processing with shock waves.

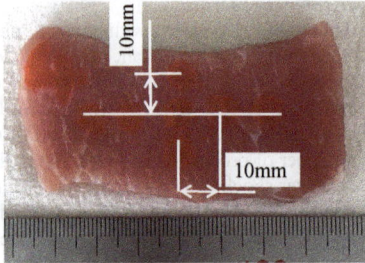

Fig. 3 Measuring points of meat (pork)

Maker: TecLock Co.,
Type: OO
Load value of spring
:203—1111mN
the shape of push needle
:hemispherical
Radius of needle: 1.19mm
Measurement duration time :1 s

Fig. 4 Specifications of durometer

2.3 Evaluation of tenderness

For the evaluation of the tenderness of meat, we mesured the tenderness of meat using a durometer. Figure 4 shows the specifications of the durometer. This measuring instrument was made in TecLock Co., the type is OO, the load value of the spring is 203-1111mN, the shape of the push needle is a hemispherical (Radius, 1.19mm), the contact time between the durometer and the meat is approximately 1 s. This measurement is in accordance with "ASTM D 2240"[26]. We measured the meat five times at each measurement point and calculated the mean.

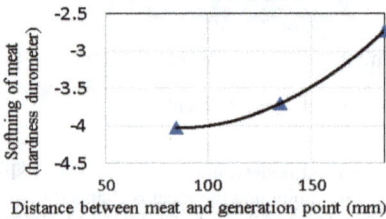

Fig.5 Relationship between distance and hardness Fig.6 Relationship between number of times of shock wave generation and hardness

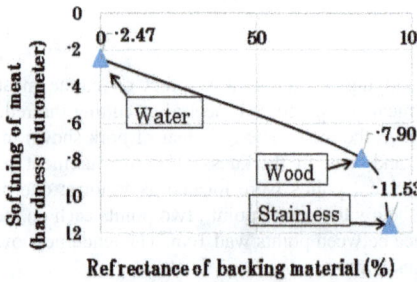

Fig.7 Relationship between backing material and hardness

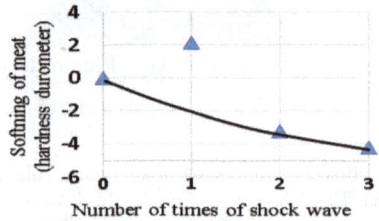

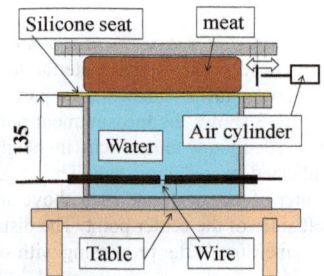

Fig.8 Picture of comcept design of pressure vessel

3. Experiment result and discussion

Figure 5 shows the relationship between the distance of the shock wave generation point from the meat (Horizontal axis) and hardness (tenderness) of meat (Vertical axis). As the distance dereases, it is obvious that the meat became tender.Figure 6 shows the relationship between number of times shock waves were generated (horizontal axis) and the hardness (tenderness) of meat (vertical axis). A negative hardness value is indicated that the meat is softened. After the third shock wave generation, meat is clearly softened by about 1.8 fold. The meat was hard after one time shock wavecompared with the other number of times of processing. The other-experimental processing (extraction and flour milling) in this study showd negative effect after one time shock wave processing[27,28]. It will be necessary clarify to the reason why the meat became hard after one-time shock wave processing, as shown by observation undar a microscope. Figure 7 shows the relationship between the reflectance of the backing material to water and tenderness.

The acoustic impedance that reflectance is decided on is calculated by the multiplication of density of the meat and the speed of sound. A different reflection expansion wave occurs by materials of the back materials, and it is assumed that a change occurred in the hardness of the meat.The effect of the backing material on tenderness was found to be larger than those of the number of times of shock wave generation and the distance between the meat and the shock wave generation point. Figure 8 shows a prototype design of the pressure vessel for processing meat using underwater shock waves. We can manufacture the pressure vessel for meat tenderness by setting the distance between the meat and the shock wave generation point, the backing material, and the transport speed of meat calculated from charging time and the numbr of times of shock wave generation according to the desired hardness (tenderness) measured using a durometer.

References

[1] Kevin Uhrmacher / NPR,National Agricultural Statistics Service,2016

[2] GLOBAL POULTRY TRENDS 2014: Poultry Meat Uptake in Europe Sure to Slow,15 October 2014

[3] S.Stoll-Kleemanna,T.O'Riordanb,The Sustainability Challenges of Our Meat and Dairy Diets,vo.57,Num.3,EnvironmEnt,2015

[4] World population prospects, the 2015 revision, United nations, Department of economic and social affairs

[5] Hiroshi Sekiguchi, Yukio Machida and Sadao Omata , Evaluation of the Hardness of Foods measured by The New Tactile Sensor for Detecting Hardness, Jpn Pediatr Soc., Vol.34,No.4,1,99-109,(1996)

[6] Takashi OKAZAKI,Kanichi SUZUKI,Shizuhiko MAESHIGE and Kiyoshi KUBOTA, Simulation of Potato Tenderness during Non-isothermal and Isothermal Processes, Nippon Shokuhin Kogyo Gakkaishi

Vol.39, No.4, 295 ~ 301 (1992), https://doi.org/10.3136/nskkk1962.39.295

[7] M B Solomon, J B Long and J S Eastridge,The hydrodyne: a new process to improve beef tenderness,Vol.75 No.6,Journal of Animal Science,p.1534-1537, https://doi.org/10.2527/1997.7561534x

[8] Tomas Bolumar, Mathias Enneking, Stefan Toepfl,Volker Heinz,New developments in shockwave technology intended for meat tenderization: Opportunities and challenges,Meat Science 95 (2013) 931–93, https://doi.org/10.1016/j.meatsci.2013.04.039

[9] Donald de Fremery,Morris F. Pool, Biochemistry of Chicken Muscle as Related to Rigor Mortis and Tenderization,Journal of Food Science,Volume 25, Issue 1,January 1960,Pages 73–87, https://doi.org/10.1111/j.1365-2621.1960.tb17938.x

[10] I.H. Hwanga, C.E. Devineb, D.L. Hopkinsc, The biochemical and physical effects of electrical stimulation on beef and sheep meat tenderness, Volume 65, Issue 2, October 2003, Pages 677–691, https://doi.org/10.1016/s0309-1740(02)00271-1

[11] Huff Lonergan E1, Zhang W, Lonergan SM., Biochemistry of postmortem muscle - lessons on mechanisms of meat tenderization,Meat Science,Volume86, Number 1,2010, pages 184-95, https://doi.org/10.1016/j.meatsci.2010.05.004

[12] Koohmaraie M.Biochemical factors regulating the toughening and tenderization processes of meat, Meat Science,Volume 43, Supplement 1, 1996, Pages 193-201, https://doi.org/10.1016/0309-1740(96)00065-4

[13] Huff Lonergan E1, Zhang W, Lonergan SM.,Biochemistry of postmortem muscle - lessons on mechanisms of meat tenderization,Meat Science,Volume 86, Issue 1, September 2010, Pages 184–195, https://doi.org/10.1016/j.meatsci.2010.05.004

[14] Anne y Castro Marques, Mário Roberto Maróstica Jr., and Gláucia Maria Pastore, Some Nutritional, Technological and Environmental Advances in the Use of Enzymes in Meat Products, SAGE-Hindawi Access to Research,Enzyme Research,Volume 2010, Article ID 480923, https://doi.org/10.4061/2010/480923

[15] I.N.A. Ashie,T.L. Sorensen,P.M. Nielsen, Effects of Papain and a Microbial Enzyme on Meat Proteins and Beef Tenderness, Journal of Food Science,Volume 67, Issue 6,August 2002,Pages 2138–2142, https://doi.org/10.1111/j.1365-2621.2002.tb09516.x

[16] D. A. King,M. E. Dikeman,T. L. Wheeler,C.L.Kastner and M. Koohmaraie, Chilling and cooking rate effects on some myofibrillar determinants of tenderness of beef, Journal of Animal Science,Volume 81,Number 6,pp.1473-1481, https://doi.org/10.2527/2003.8161473x

[17] B. B. MaARSH, N. G. LEET,M. R. DICKSON, The ultrastructure and tenderness of highly cold-shortened muscle, Journal of Food Science and Technology,Volume 9,Issue 2,June 1974,Pages 141–147, https://doi.org/10.1111/j.1365-2621.1974.tb01757.x

[18] B. B. MARSH,N. G. LEET, Studies in Meat Tenderness. III. The Effects of Cold Shortening on Tenderness, Journal of Food Science and TechnologyVolume 31, Issue 3,May 1966,Pages 450–459, https://doi.org/10.1111/j.1365-2621.1966.tb00520.x

[19] S. Itoh, K. Hokamoto, Explosion, Shock Wave and Hypervelocity Phenomena in Materials II, Materials Science Forum (Volume 566), November 2007,pp. 361-372, https://doi.org/10.4028/0-87849-465-0.361

[20] Yoshikazu HIGA, Hirofumi IYAMA, Ken SHIMOJIMA, Atsushi YASUDA, Osamu HIGA, Ayumi TAKEMOTO and Shigeru ITOH, Computational Simulation for Evaluation of Food Tenderness Treatment Vessel using Underwater Shockwave, ASME 2016 Pressure Vessels & Piping Division Conference PVP2016 PVP2016-63530 Aug, 2016, https://doi.org/10.1115/pvp2016-63530

[21] Ken SHIMOJIMA, Yoshikazu HIGA, Osamu HIGA, Ayumi TAKEMOTO, Hirofumi IYAMA, Atsushi YASUDA, Toshiaki WATANABE and Shigeru ITOH, Visualization of Shock Wave Propagation Behavior of the General-Purpose Batch Processing for Pressure Vessel by Numerical Simulation, ASME 2016 Pressure Vessels & Piping Division Conference PVP2016 PVP2016-63510 ,Aug,2016, https://doi.org/10.1115/pvp2016-63510

[22] Ken Shimojima,Yoshikazu Higa,Osamu Higa,Atsushi Yasuda,Ayumi Takemoto,Shigeru Itoh,Hirofumi Iyama,Toshiaki Watanabe,Development of Prototype Machine for Food Processing by Underwater Shock Wave, The 2nd International Conference on Engineering Science and Innovative(ESIT 2016), Phuket, Thailand, April 21 - 23 (2016), PaperID:No.12,PP.360-365, https://doi.org/10.1115/pvp2016-63530

[23] Yoshikazu HIGA,Hirofumi IYAMA,Ken SHIMOJIMA,Atsushi YASUDA,Osamu HIGA, Ayumi TAKEMOTO,Shigeru ITOH,Computational Simulation of Shock Wave Propagation in Foods,The 2nd International Conference on Engineering Science and Innovative(ESIT 2016), Phuket, Thailand, April 21 - 23 (2016),PaperID:No.13,PP.366-371, https://doi.org/10.1109/esit.2018.8665029

[24] Higa, K. Higa, H. Maehara, S. Tanaka, K. Shimojima, A. Takemoto, K. Hokamoto and S. Itoh; EFFECTS OF IMPROVING CURRENT CHARACTERISTICS OF SPARK DISCHARGE ON UNDERWATER SHOCK WAVES, The International Journal of Multiphysics, Vol. 8, No. 2, pp.245-252, 2014, https://doi.org/10.1260/1750-9548.8.2.245

[25] Ben-Dor, Gabi,shock Wave Reflection PhenomenaSpringer,Springer

[26] Standard Test Method for Rubber Property—Durometer Hardness, Active Standard ASTM D2240-15

[27] Ken SHIMOJIMA, Osamu HIGA, Yoshikazu HIGA, Ayumi TAKEMOTO, Hirofumi IYAMA, Atsushi YASUDA, Toshiaki WATANABE and Shigeru ITOH,Production of Rice Powder Milling Flour Device and Characterization by Numerical Simulation,ASME 2016 Pressure Vessels & Piping Division Conference PVP2016 PVP2016-63588 Aug.2016, https://doi.org/10.1115/pvp2016-63588

[28] K. Shimojima, A. Takemoto, M. Vesenjak, Y. Higa, Z. Ren, S. Itoh,The effect of improving the oil extraction of Slovenia production seed by underwater shock wave,MULTIPHYSICS 2015,10-11 Dec 2015, London, United Kingdom,P.9

Explosion Shock Waves and High Strain Rate Phenomena
Materials Research Proceedings 13 (2019) 41-46

Materials Research Forum LLC
https://doi.org/10.21741/9781644900338-7

Deformation Behavior of a Polygonal Tube under Oblique Impact Loading

Yohei SHINSHI[1,a], Makoto MIYAZAKI[1,b*] and Keisuke YOKOYA[2,c]

[1] National Institute of Technology, Nagano College, 716 Tokuma, Nagano-city, Nagano 381-8550, Japan

[2] Tokyo Institute of Technology, 2-12-1 Ookayama, Meguro-ku, Tokyo 152-8550, Japan

[a]syourunner1920@gmail.com, [b]miyazaki@nagano-nct.ac.jp, [c]powerto84@gmail.com

Keywords: Dynamic Deformation, Impact Load, Plastic Buckling, Numerical Analysis

Abstract. Aluminum tubes are energy-efficient absorbing components and are widely used for framework and reinforcement materials of structures. The effects of the axial length and cross-sectional shape on the deformation behavior were investigated. Regarding the axial length, it has changed only to a certain length, and there are few studies on it. This paper deals with the influence of axial length. Also, when an impact is actually applied to the square tube, the impact in the oblique direction must also be taken into consideration. Therefore, the deformation behavior was analyzed by applying impact to the square tube from various angles other than the axial direction. An analysis of the dynamic deformation process of the polygonal tube was made using a finite element method. The results show that the load reached the peak immediately after the weight hit the square tube, then declined gently. The same tendency was obtained even if the axial length was changed. However, as the axial length became longer, the displacement taken to reach the peak load increased. As for the impact in the oblique direction, the peak load was small as compared with the axial direction. The deformation of square tube did not buckle in whole but only partially at any length.

Introduction

Square tubes have been used for framework and reinforcement members of structures.There are many studies on circular tubes, and deformation behaviors have been studied by static and dynamic compression tests [1]. Previous studies have shown that square tubes have a role of absorbing impact energy by crushing under pressure in the axial direction at the time of a collision [2]. Aluminum alloy has a Young's modulus that is one-third that of commonly used steel materials, giving it the disadvantage of low rigidity. In addition, the whole buckles become large when thickness is increased, and causing axial compression deformation, which cannot effectively absorb collision energy[3].The tubular bodies with polygonal tubes and cellular cross sections have been studied as a means to effectively absorb energy [4]. Additionally, an influence of axial length on dynamic axially compressed aluminum tubes is being considered [5-7].It is known that elastic deformation occurs in the entire square tube prior to plastic deformation when the square tube deforms. Since this is periodic and wavy, it seems that the axial length will have a large influence. In a previous study, deformation behaviors up to 500 mm in length have been considered [8]. The purpose of this paper is to discuss, the deformation behavior of dynamic axial compression of an aluminum square tube of axial lengths of 500 mm, 750mm and 1000 mm. Also, when an impact is applied to the tube, the impact in the oblique direction must also be taken into consideration. Therefore, for comparison with the axial compression, deformation behavior of aluminum square tube under oblique impact loading was considered.

Explosion Shock Waves and High Strain Rate Phenomena
Materials Research Proceedings **13** (2019) 41-46

Materials Research Forum LLC
https://doi.org/10.21741/9781644900338-7

Numerical Analysis

Analytical method. The analysis is conducted by non-linear structure analysis program (Marc 2018) and pre-post processor (Mentat 2018). An example of analytical model is shown in Fig. 1. The specimen is an aluminum tube (A6063-T5). Material properties are shown in Table 1. Concerning axial length l, the square tubes (l = 500 mm, 750 mm and 1000 mm) are discretized at 20000, 30000, 40000 bilinear four-node shell elements, respectively. For the analysis in the oblique direction, an angle of θ= 10 degrees was given between the weight and the impact edge of the square tube. Schematic diagram of the analysis model is shown in Fig. 2.

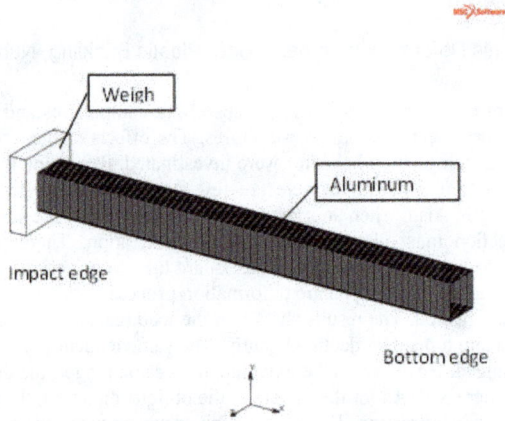

Fig. 1 Analytical model of square tube (l = 500mm).

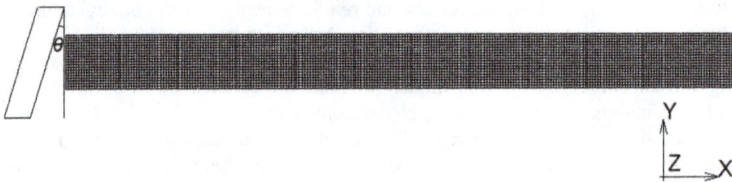

Fig. 2 Impact angle of weight (θ= 10 deg).

The nodes on the edge of the tube are fixed with the exception of in the axial direction of the impact edge. The weight (80 × 80 × 20 mm, 15 kg) is an un-discretized three-dimensional, eight-node, first-order, isoparametric element. The deformed tube is regarded as an isotropic material following von-Mises yield condition, and the flow stress-strain relationship is shown in Equation (1) because the effect of the strain rate of the aluminum is smaller than that of other materials like iron, etc [9].

Explosion Shock Waves and High Strain Rate Phenomena
Materials Research Proceedings 13 (2019) 41-46

Materials Research Forum LLC
https://doi.org/10.21741/9781644900338-7

Table1 Material properties of aluminum (A6063-T5)

Young's modulus	E [GPa]	69
Poisson's ratio	v	0.33
Density	ρ [kg/m^3]	2.71×10^3
Work-hardening modulus	F[MPa]	268
Work-hardening exponent	n	0.065

$$\sigma = F\varepsilon^n. \qquad (1)$$

In this analysis, the time step width is 1μs. The Newton-Raphson method and the updated Lagrangian formulation are used as the solution methods of the non-linear equation, and the single-step houbolt of implicit solution time-integration method is used for the analysis of dynamic deformation. The impact velocity is 10 m/s.

Results and Considerations

Analytical results. Final deformations of the square tube (l = 500 mm, 750 mm and 1000 mm)are shown in Fig. 3.

(a)   (b)

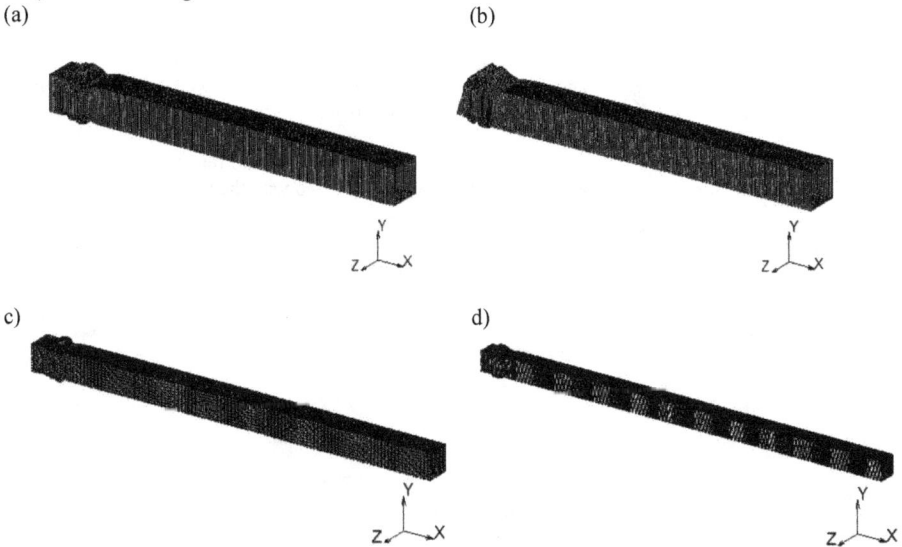

c)   d)

Fig. 3 Final deformations of square tube (a)l = 500 mm, (b)l = 500 mm (oblique direction), c)l = 750 mm, (d)l = 1000 mm.

Load-displacement curve. Load-axial displacement curves of square tube with length variation are shown in Fig. 4. The same tendency was obtained even if the axial length was changed. The peak load is not hardly affected by axial length. However, it was confirmed that the displacement taken to reach the peak load is dependent on the axial length. And the final displacement is independent of the axial length.

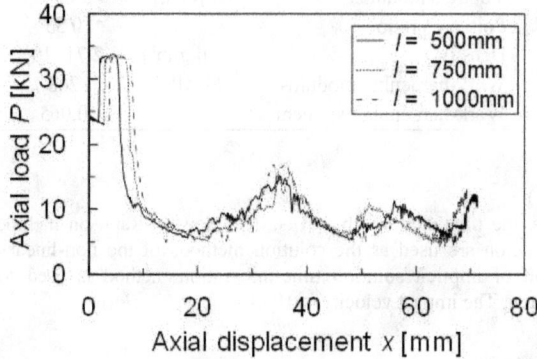

Fig. 4 Relationship between axial load P *and axial displacement x (axial loading,l = 500 mm, 750 mm and 1000 mm).*

The load-displacement curves in the axial direction and oblique direction of length l = 500 mm are shown in the Fig. 5.Compared with the result of loading in the axial direction, the peak load is small. And the displacement taken to reach the peak load is larger than that in the axial direction.This is because the weight collides with the upper portion of the tube so that the initial collision portion is smaller than the axial direction. After that, the load reached a peak at around 17 mm in axial displacement, because the entire weight collided with the tube.

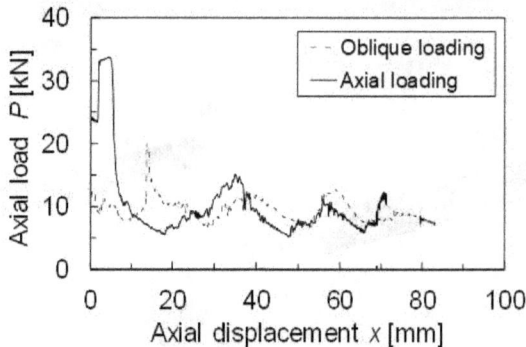

Fig. 5 Relationship between axial load P *and axial displacement x (axial loadingand oblique loading,l = 500mm).*

Effects of axial length on compressive strain distribution. Strain distributions in bucking region are shown in Fig. 6. As seen in Fig. 6, it was confirmed that peak strain wasnot considerably affected by changes in the axial length. The strain occurs partially at the side of the tube even if the axial length changes. In addition, the strain is almost 0 except for the peak strain region. From the above results, it was also confirmed that deformation occurred partially.

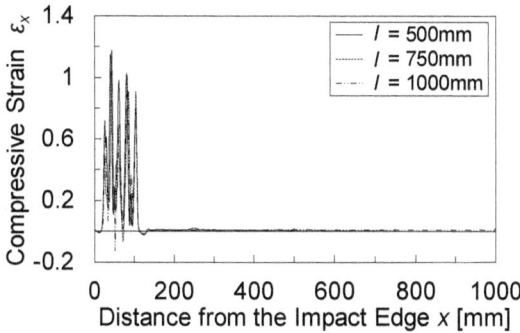

Fig. 6 Compressive strain distribution along the side.

Effects of axial length on axial strain distribution. Strain distributions along the hill and valley are shown in Fig. 7. The distortion here shows the compressive strain at the position where the concave-convex pattern of the edge is maximized.It was confirmed that axial strain near the corner of the tube is large compared to that of the other parts of tube. Even if the axial length changes, the trends of the axial strain distribution was not considerably affected. In addition, the peak strain did not markedly change.

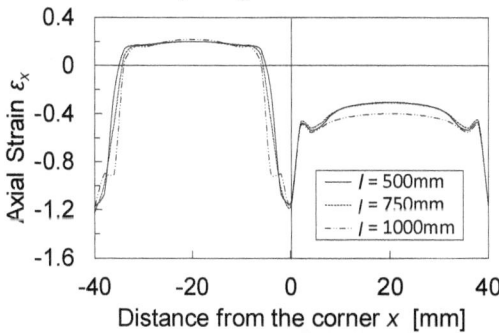

Fig. 7 Axial strain distribution along the hill and valley.

Conclusions

The deformation behavior of the square tube subjected to dynamic axial compression was compared with the results of examination by finite element analysis with the axial length varying from 500 mm to 1000 mm and the deformation behavior in the axial direction and diagonal direction when the axial length was 500 mm. The following conclusions were obtained.

(1) The trends of the load-displacement curve and peak load were not considerably affected by changes in the axial length.

(2) As the axial length increases, the final displacement becomes smaller.
(3) In the case of oblique direction impact, it was found that the peak load is small compared tothat of axial direction impact, and the displacement taken to reach the peak load is dependent on the axial length.
(4) The deformation of the square tube did not buckle in whole but only partially.

References

[1] W. Abramowicz and N. Jones, International Journal Impact Engineering. **2**-2 (1984) 179-208.

[2] M. Yamashita, H. Kenmotsu and T. Hattori : Thin - Walled Struct., **69** (2013), 45-53.

[3] D-K. Kim, S. Lee and M. Rhee, Materials & Design. **19**-4 (1998) 179-185.

[4] J Fang, Y. Gao, G. Sun, N. Qiu . and Q. Li : Thin-Walled Struct., **95** (2015), 115-126.

[5] M. Miyazaki, H. Endo, and H. Negishi, J. Mater. Process. Technol., **85**-1-3 (1999) 213-216.

[6] M. Miyazaki and H. Negishi, Materials Transactions. **44**-8 (2003) 1566-1570.

[7] M. Miyazaki and M Yamaguchi, Procedia Engineering. **81** (2014) 1067-1072.

[8] K. Yokoya, M. Miyazaki, Y. Tojo and M. Yamashita : Procedia Eng., **207** (2017), 251-256

[9] S. Tanimura, H. Hayashi, T.Yamamoto and K. Mimura, J. Solid Mech. & Mater., **3**-12 (2009), 1263-1273.

Explosion Shock Waves and High Strain Rate Phenomena   Materials Research Forum LLC
Materials Research Proceedings 13 (2019) 47-52   https://doi.org/10.21741/9781644900338-8

Collision Behavior in Magnetic Pressure Parallel Seam Welding of Aluminum Sheets

Akira HATTA[1,a], Makoto MIYAZAKI[1,b*] and Yohei KAJIRO[1,c]

[1]Department of Mechanical Engineering, National Institute of Technology, Nagano College, 716 Tokuma, Nagano-city, Nagano 381-8550, Japan

[a]5030arika@gmail.com,[b]miyazaki@nagano-nct.ac.jp, [c]kajiro.nagano@gmail.com

Keywords: Magnetic Pressure Seam Welding, Parallel Seam Welding, Collision Behavior, Aluminum Sheet, Numerical Analysis

Abstract. Magnetic pressure seam welding has attracted attention as a new joining method for aluminum thin plates. Magnetic pressure seam welding is a collision welding process, utilizing electromagnetic force as the acceleration mechanism. The electromagnetic seam welding is a method of abruptly adding a high density magnetic flux around a metal material and utilizing the generated electromagnetic force to deform the thin plate at high speed and pressure welding. This paper deal with the deformation behavior of parallel aluminum seam welded aluminum sheet. Numerical analysis of the dynamic deformation process of the metal plate is performed by the finite element method. The sample used for this analysis is assumed to be a thin plate made of aluminum (A1050-H24, width100mm, thickness 1mm) and composed of quadrilateral elements of plane strain. The experimental results show that the collision speed between the aluminum plates is sufficiently reproduced. The impact point velocity between the aluminum plate surfaces was very high at the initial collision point but decreased continuously during welding. It was also found that the smaller the gap is, the faster the collision point moving speed becomes.

Introduction

Aluminum has higher electrical conductivity and thermal conductivity than iron, so welding is difficult due to low heating efficiency. In previous studies, there is a report on the magnetic pressure seam welding method [1]-[14]. Magnetic pressure seam welding is a collision welding process similar to explosive welding and utilizes electromagnetic force as an acceleration mechanism. Magnetic pressure seam welding accelerates and collides a certain metal plate (flyer plate) to another stationary metal plate (parent plate) by using electromagnetic force. When an impulse current from a capacitor bank passes through a flat one-turn coil, a magnetic flux is instantaneously generated in the coil. The eddy currents are induced in insulated flyer plate in the coil. In magnetic pressure parallel seam welding, one-turn coils are arranged in parallel. A part of flyer plate along the longitudinal direction of the coil bulged toward a parent plate, then flyer plate collided and was welded to a parent plate. At the time of the high-speed collision, metal jets are emitted in the welding interface of the specimen [7]. The collision point velocity and collision angle are determined by the primary and induced electromagnetic force. True metallic bonding is achieved at the mating interface if contact takes place above an appropriate collision point velocity and collision angle [15]. The purpose of this paper is to discuss, the dynamic deformation behavior of magnetic pressure parallel seam welding of aluminum sheets.

Welding principle

The welding principle is shown in Fig. 1. Magnetic pressure parallel seam welding uses electromagnetic force to accelerate one metal sheet (flyer plate) against another stationary metal

Explosion Shock Waves and High Strain Rate Phenomena   Materials Research Forum LLC
Materials Research Proceedings **13** (2019) 47-52   https://doi.org/10.21741/9781644900338-8

sheet (parent plate). When a high magnetic field **B** suddenly occurs and enters the metal sheet, eddy current (current density **i**) passes through the metal sheet. As a result, the electromagnetic force of Eq. 2 acts mainly on the flyer plate and it is accelerated away from the coil and collides rapidly with the parent plate [10]. The eddy current **i**, electromagnetic force **f** and Joule heat Q are given as follows. κ and **B** are electric conductivity and magnetic flux density at aluminum sheet. When the residual inductance of the electromagnetic forming apparatus is large, it becomes difficult for a large current to flow through the one-turn coil, so the magnetic pressure also becomes small and it is difficult to join. Since the inductance of the coil of the multi-turn coil is higher than that of the one-turn coil, the current flowing through the coil can be increased and the magnetic pressure can be increased.

$$\text{rot } \mathbf{i} = -\kappa \frac{\partial \mathbf{B}}{\partial t} \tag{1}$$

$$\mathbf{f} = \mathbf{i} \times \mathbf{B} \tag{2}$$

$$Q = \frac{i^2}{\kappa} \tag{3}$$

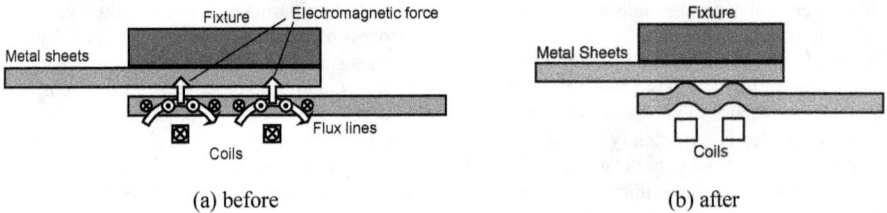

(a) before   (b) after

Fig. 1 Schematic illustrations of welding method

Numerical Analysis

The analysis was made by non-linear-structure-analysis-program (MARC 2018). An element division model and boundary condition are shown in Fig. 2. In this analysis, the metal sheets (100mm width, 1mm thickness) were assumed to be composed of 20000 plane-strain quadrilateral elements. The gap length between the flyer plate and the parent plate was varied between from 0.5mm to 1.5mm.The deformed plate was an isotropic material. The true stress-true strain relation is given by Eq. 4.

$$\sigma = F\varepsilon^n \tag{4}$$

The specimens used in this simulation were aluminum sheets. Material properties obtained by a static tension test are shown in Table 1. Okagawa et al. reported that deformation of aluminum sheets was finished at the first discharge waveform [16]. In the analysis, calculation time was 10.8 μs, and calculation step was 5000. The time integration method was single-step houbolt of implicit solution method. The magnetic pressure P - measured magnetic flux density B relations are given by Eq. 5.

$$P = \frac{B}{2\mu}\left\{1 - \exp\left(-\frac{2t}{\delta}\right)\right\} \tag{5}$$

Explosion Shock Waves and High Strain Rate Phenomena　　　　　Materials Research Forum LLC
Materials Research Proceedings **13** (2019) 47-52　　　　　https://doi.org/10.21741/9781644900338-8

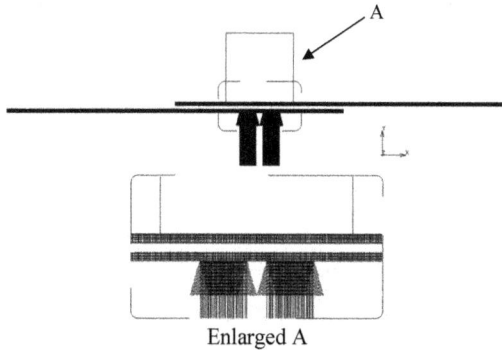

Enlarged A

Fig. 2 Finite element model of magnetic pressure paralell seam welding.

Table 1 Material properties of A1050-H24 sheets

Young's modulus	E [GPa]	69
Poisson's ratio	ν	0.33
Density	ρ [kg/m^3]	2.71×10^3
Strength coefficient	F [MPa]	118
Strain hardening exponent	n	0.0623

The parameters μ, δ and t are magnetic permeability, skin depth and thickness of metal sheets, respectively. The relationship between time and magnetic pressure is shown in Fig. 3.

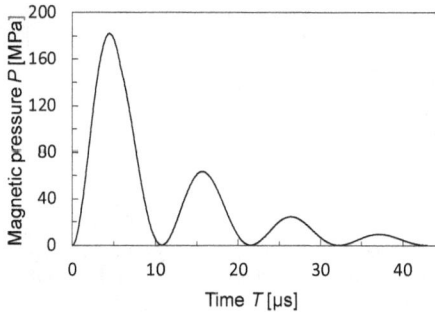

Fig. 3 Relationship between time T and magnetic pressure P.

Results and considerations
The analysis in this study was examined from the initial collision point to the outside.

Collision angle. The collision angle and collision point velocity relations[15] are given as follows:

$$v_p = 2v_c \sin\frac{\beta}{2} \tag{6}$$

$$v_p = \sqrt{v_x{}^2 + v_y{}^2} \tag{7}$$

Where β, v_c, vp, v_x and v_y are collision angle, collision point velocity, collision velocity, x direction element of v_p and y direction element of v_p, respectively. Spatial relationships of v_p, v_c and β are shown in Fig. 4.

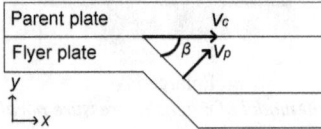

Fig. 4 Definition of collision angle β, collision velocity v_p and welding velocity v_c.

The relation between distance from initial collision point and collision angle is shown in Fig. 5. Watanabe et al. reported that the collision angle between the metal plate surfaces was 0 degree at the initial collision point [17]. In this simulation, the collision angle between the metal plate surfaces was 0 degree at the initial collision point, but it increased continuously during the welding. Also, increasing the gap length increased the maximum value of the collision angle. When the gap length was 1.5 mm, the collision angle sharply increased at about 2 mm from the initial collision point.

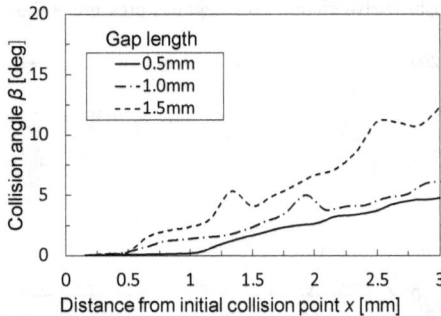

Fig. 5 Relationship between distance from initial collision point x and collision angle β.

Collision velocity. The relation between initial collision point and collision velocity is shown in Fig. 6. Depending on the gap length, the collision speed is fast. The collision speed became the maximum value at a certain distance away from the initial collision point and decelerated from there. This phenomenon occurs when the movable plate continues to be accelerated by electromagnetic force. The collision speed varies with the electromagnetic force and the mechanical characteristics of the movable plate.

Explosion Shock Waves and High Strain Rate Phenomena | Materials Research Forum LLC
Materials Research Proceedings **13** (2019) 47-52 | https://doi.org/10.21741/9781644900338-8

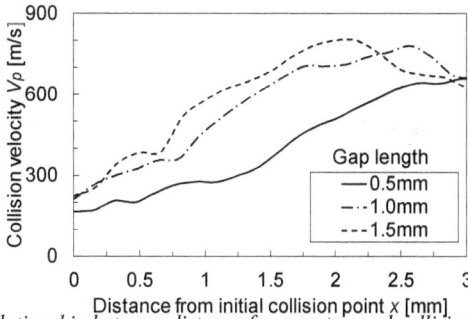

Fig. 6 Relationship between distance from centerxand collision velocity v_p.

Collision point velocity. The relation between initial collision point and collision point velocity is shown in Fig. 7. The collision point velocity was very high-speed at the initial collision point, but it decreased continuously during the welding. As the gap length increases, the initial collision point movement speed in the vicinity of the initial collision point also increases. When surface layer of the metal plate is emitted as a metal jet, strong metallic bond is established on the metal plate [18]. The velocity of metal jet is 2000-3000 m/s [19]. As seen in Fig. 7, the collision point velocity of near the initial collision point is higher than velocity of metal jet. The metal jet is not emitted near the initial collision point. Therefore, the un-welded zone exist near the initial collision point.

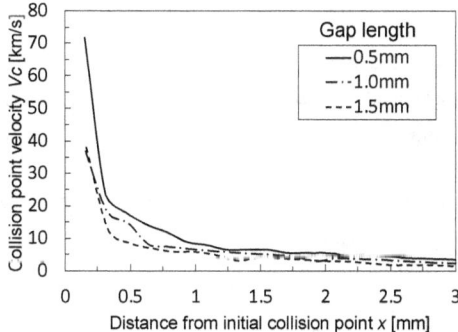

Fig. 7Relationship between distance from initial collision point x and collision point velocity v_c.

Conclusions

(1) The collision point velocity is very high-speed at the initial collision point, but it decreases continuously during the welding.

(2) The collision angle between the metal plate surfaces is 0 degree at the initial collision point, but it increases continuously during the welding.

(3) As the gap length increases, the maximum value of the collision angle increases.

(4) The maximum value of the crash velocity tends to increase as the gap length increases.

(5) The analysis from the initial collision point to the outer side is similar to the single coil.

Acknowledgment
This work was supported by a Grant-in-Aid for Scientific Research (C) (16K06757) from Japan Society for the Promotion of Science (JSPS).

References
[1] T. Aizawa, K. Okagawa, M. Yoshizawa and N. Henmi: Proc. of 4th Int. Symp. on Impact Engineering (2001) 827-832.

[2] T. Aizawa: J. of Japan Inst. of Light Metals, 54 (2004) 153-158.

[3] K. Okagawa and T. Aizawa: J. Jpn. Soc. Technol. Plast., 48 (2007) 323-327.

[4] M. Watanabe, S. Kumai and T. Aizawa: Mater. Sci. Forum, 519-521 (2006) 1145-1150.

[5] K. J. Lee, S. Kumai, T. Arai and T. Aizawa: Mater. Sci. Eng. A, 471 (2007) 95-101.

[6] T. Aizawa, M. Kashani and K. Okagawa: Weld. J., 86 (2007) 119s-124s.

[7] M. Watanabe and S. Kumai: Mater. Trans., 50 (2009) 2035-2042.

[8] S.D. Kore, P.P.Date and S.V. Kulkarni: Int. J. of Impact Eng., 34 (2007) 1327-1341.

[9] H. Serizawa, I. Shibahara, S. Rashed and H. Murakawa: Mater. Sci. Forum, 638-642 (2010) 2166-2171.

[10] T. Aizawa, K. Okagawa, M. Kashani: J. Mater. Process. Technol, **213**-7 (2013) 1095–1102.

[11] A.Stern, O. Becher, A.Stern, M. Nahmany, D. Ashkenazi, V. Shribman: Weld. J., 94 (2015) 257S-264S.

[12] T. Aizawa, K. Matsuzawa: J. of Jpn. Weld. Soc., 33 (2015) 130s-134s.

[13] M. Watanabe, S. Kumai, K. Okagawa, T. Aizawa: Aluminium Alloys, 2 (2008) 1992-1997.

[14] M. Miyazaki, K. Sasaki and M. Okada: Mater. Sci. Forum, 767 (2014) 166-170.

[15] K. Hokamoto, M. Fujita, M. Ohtsuka: Impact Forming (High-energy Rate Forming), CORONA Publishing Inc., Tokyo, (2017) 66-74.

[16] K. Okagawa and T. Aizawa: J. Jpn. Soc. Technol. Plast., 47 (2006) 632-636.

[17] M. Watanabe, S. Kumai, K. Okagawa, T. Aizawa: Pre-Prints of the 82nd National Meeting of JWS, (2008) 122-123.

[18] M. Watanabe and S. Kumai: Mater. Trans., 50 (2009) 286-292.

[19] S. Kakizaki, M. Watanabe and S. Kumai: Proc. of the 12th Int. Conf. on Aluminium Alloys (2010) 945-949.

[20] K. Okagawa, M. Ishibashi, E. Kabasawa and H. Yamagishi: Proc. of the 66th Jpn. Joint Conf. for the Technol. of Plast., (2015) 389-390.

[21] S. Kakizaki, M. Watanabe and S. Kumai: Mater. Trans., 52 (2011) 1003-1008.

Explosion Shock Waves and High Strain Rate Phenomena   Materials Research Forum LLC
Materials Research Proceedings 13 (2019) 53-56   https://doi.org/10.21741/9781644900338-9

Azimuthal Characteristics on Blast Wave from a Cylindrical Charge – Small Scale Experiment –

Tomotaka Homae[1, a *],Yuta Sugitana[2,b],Tohoharu Matsumura[2,c]and Kunihiko Wakabayashi[2,d]

[1]Department of Maritime Technology, National Institute of Technology, Toyama College,1-2 Ebie-neriya, Imizu, Toyama 933-0293 JAPAN

[2] Research Institute of Science for Safety and Sustainability, National Institute of Advanced Industrial Science and Technology, Central 5, 1-1-1 Higashi, Tsukuba, Ibaraki 305-8565 JAPAN

[a]homae@nc-toyama.ac.jp, [b]yuta.sugiyama@aist.go.jp, [c]t-matsumura@aist.go.jp, [d]k-wakabayashi@aist.go.jp

Keywords: Explosion, Azimuth Angle, Non-Spherical Explosive, Blast Pressure, Impulse

Abstract. Azimuthal characteristics on blast wave from a cylindrical charge were experimentally investigated. A cylindrical PETN pellet, weight of 0.50 g, was detonated on a large steel plate, which is a model of the ground surface. 0°was defined as the direction of detonation, which was correspondence with the central symmetry axis of the cylindrical pellet. 12 pressure transducers were embedded in the steel plate to measure the pressure histories on the plate. The direction of the pellet and detonation was rotated every 30° and the distribution of blast pressure histories around the explosive were obtained. The peak overpressure and impulse were high in the range from 30° to 80° compared with the standard explosion data in which the explosive was placed vertically for a two-dimensional axisymmetric explosion on the steel plate and detonated from the top of the explosive. On the contrary, these blast parameters were low in the range from 90° to 130°. These blast pressures were not low in the direction of 180°. These findings are important for safety. The data will be compared with numerical simulations in future.

Introduction

Explosion of non-spherical explosive causes anisotropic blast wave. This anisotropic blast waves have been extensively studied for many decades. Especially, explosion of cylindrical explosive, which is relatively symmetric, have been studied both experimentally and numerically. R.A. Strehlow and W.E. Baker reviewed these studies in 1976 [1]. As the experimental technique and numerical analysis technique have been improving, many papers have been published until present [2]. It is noteworthy for safety that the blast wave in specific direction is reported to be strong in these papers.

The authors carried out indoor tabletop experiments for evaluating the blast wave, using explosive of 1 g scale, and the obtained data were examined by numerical analysis. The authors reported the blast wave mitigation or distribution in a couple of systems [3-6]. In this study, the authors applied the technique above to evaluate the blast wave distribution from a cylindrical explosive precisely. The experimental system was designed under consideration for numerical analysis.

Experiment

Test Explosives. A PETN pellet, weight of 0.50 g, was used as a test explosive. Its length and diameter were both 7.5 mm. It contained 5 wt% of carbon for forming. A specially designed electric detonator with 100 mg lead azide was used as a detonator. Both the PETN pellet and the detonator were distributed by Showa Kinzoku Kogyo Co. Ltd. A spacer was used so as to the height of the center of the explosive was 0.18 m/kg$^{1/3}$. The scaled height was set up as the same-scaled height of previous study [5] for comparison. The spacer was made of pasteboard and was rectangular block with the height of 10 mm and base of 7x7 mm. The lateral surface of the cylindrical pellet was fixed on the spacer using epoxy resin adhesive. 4 kV was applied to initiate the detonator (test explosive) using a firing system (FS-43; Teledyne RISI, Inc.).

Ground Surface Model. A steel plate, length of 3510 mm, width of 2200 mm, and thickness of 10 mm, was regarded as a ground surface. The test explosive with the spacer was fixed on the surface model using double-sided adhesive tape. Fig. 1 shows the configuration of the test explosive on the surface model. The direction of ignition is defined as 0°.

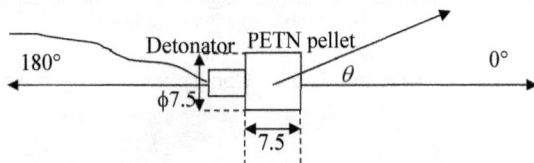

Fig. 1 Set up of test explosiveand definition of the angle.

Pressure Measurement. 12 pressure transducers (113B28; PCB Piezotronics, Inc.) were used to measure the blast pressure at the ground model surface. The pressure transducers were set with the vibration isolator (GEL Tape; Taica Corporation) as the diaphragm of the transducer was flush with the ground surface model. The transducers were on three lines of every 10°. Then, the pressure histories of three directions, for example, 0°, 10° and 20°, were obtained in one explosion experiment. The distances from the center of the explosive were 400 mm, 800 mm, 1200 mm, and 1600 mm, respectively. The corresponded Hopkinson scaled distance was from 5.1 m/kg$^{1/3}$ to 20.4 m/kg$^{1/3}$. The Hopkinson scaled distance was obtained the distance divided by cube root of the net weight of PETN, 95% of the measured weight of the pellets. The output signals were recorded using a transient recorder (LTT184/8;LabortechnikTasler GmbH; sampling rate of 1.04 MHz and resolution of 16 bits in this study) through an amplifier system (30510 and 30622; H-Tech Laboratories, Inc.).

Number of Experiments and Standard Experiments. The setup of explosive direction varied every30°. Seven experiments gave one series of the data from 0° to 180°. Two series of the data were acquired in this study. In addition to the experiments above, the standard data in which the explosive was placed vertically for a two-dimensional axisymmetric explosion on the steel plate and detonated from the top of the explosive were obtained for comparison.

Results and Discussions
The obtained pressure histories were fitted using spline functions. Then, peak overpressures and positive scaled impulse, just scaled impulse hereinafter, were determined. Fig. 2 (a) shows the relation between peak overpressure and azimuthal angle. The shown data is average of data by two experiments. The dotted line is the peak overpressure of the standard data. Fig. 2 (a) shows the absolute value, and the dependence on azimuth angle is not easy to understand at further points. Fig. 2(b) shows the peak overpressure ratio to the standard data. It is clearly seen that the explosive shape affected even at 20.4 m/kg$^{1/3}$. The peak overpressure is high from 30° to 80° (at 5.1 m/kg$^{1/3}$, from 30° to 60°), compared with the standard data. The difference is from 10% to

Explosion Shock Waves and High Strain Rate Phenomena
Materials Research Proceedings 13 (2019) 53-56

Materials Research Forum LLC
https://doi.org/10.21741/9781644900338-9

25%. On the contrary, It is low from 90° to 120°. The peak overpressure was *not* low at the direction of 180°, that is, opposite to the direction of ignition.

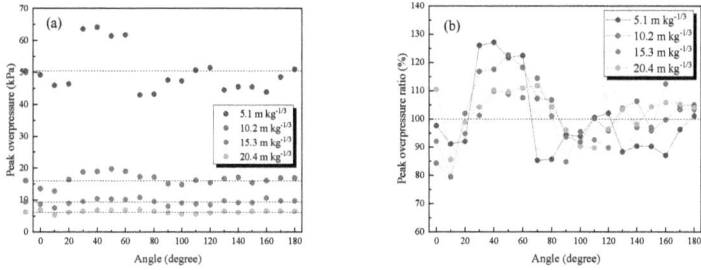

Fig. 2 (a) Dependence of peak overpressure on azimuthal angle. (b) Dependence of peak overpressure ratio to standard peak overpressure on azimuthal angle.

Fig. 3 (a) shows the dependence of scaled impulse on azimuthal angle. Fig. 3 (b) shows the dependence of the ratio on azimuthal angle. The tendency is similar to that of peak overpressure, although data scatter at 20.4 m/kg$^{1/3}$is large.

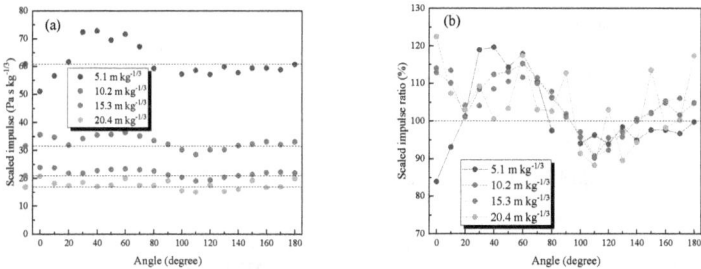

Fig. 3 Dependence of scaled impulse on azimuthal angle. (b) Dependence of scaled impulse ratio to standard scaled impulse on azimuthal angle.

The relation between peak overpressure and the scaled distance on log-log plane was fitted using quadratic function. The peak overpressures of standard data at scaled distance of 8 m/kg$^{1/3}$and 16 m/kg$^{1/3}$ were calculated using the fitted quadratic function. Scaled distance of 8 m/kg$^{1/3}$and 16 m/kg$^{1/3}$ are regulated as safety distance in law in Japan. The calculated peak overpressures were 23.67 kPa and 8.59 kPa, respectively. Then, the scaled distances, where the peak overpressures were 23.67 kPa and 8.59 kPa, were determined for every direction (from 0° to 180°). Same procedure was adopted to the scaled impulse. The scaled impulses at 8 m/kg$^{1/3}$and 16 m/kg$^{1/3}$ were 39.12 Pa s/kg$^{1/3}$ and 20.63 Pa s/kg$^{1/3}$, respectively. Figs. 4 (a) and (b) show the isobaric line of peak overpressure and scaled impulse. The plots form 190° to 350°in the figure are just the mirror image of 170° to 10°.

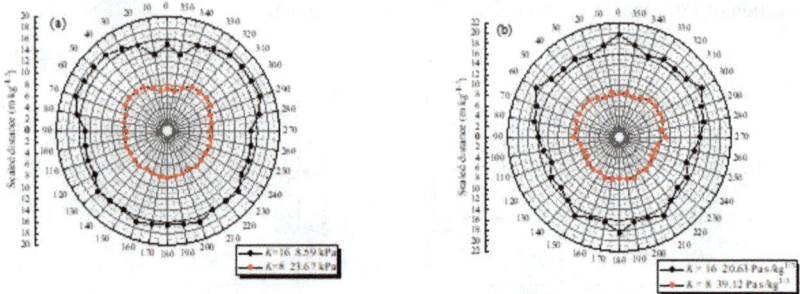

Fig. 4 Isobaric line of (a) peak overpressure and (b)scaled impulse.

Regarding to the peak overpressure, the isobaric line expanded, up to scaled distance of approximately 2 m/kg$^{1/3}$, at the direction from 30° to 80°. This means that the peak overpressure is same as the standard data at scaled ditance of 16 m/kg$^{1/3}$, at scaled distance of 17 m/kg$^{1/3}$ or even 18 m/kg$^{1/3}$ on the specific direction. Regarding to the scaled impulse, the tendency is similar to that of peak overpressure, however, the range of angle that expanded is larger than that of peak overpressrure. Both isobaric lines are not symmetric on the line of 90° and 270°. The blast wave is strong at former direction. The direction of ignition might affect the distribution.

In conclusion, the shape of the explosive and direction of ignition affected the blast pressure distribution. The blast pressure from 30° to 80° were high. These factsare important for safety consideration.

References

[1] R.A. Strehlow and W.E. Baker,The characterization and evaluation of accidental explosions,Prog. Energy Combut. Sci., 2 (1976) 27-60.

[2] H. Kleine, J. M. Dewey, K. Ohashi, T. Mizukaki and K. Takayama, Studies of the TNT equivalence of silver azide charges, Shock Waves, 13 (2003) 123-138. https://doi.org/10.1007/s00193-003-0204-3

[3] T. Homae, Y. Sugiyama, K. Wakabayashi, T. Matsumura, and Y. Nakayama, Blast wave mitigation from the straight tube by using water Part I -Small scale experiment-, Materials Sci. Forum, 910 (2018) 149-154. https://doi.org/10.4028/www.scientific.net/msf.910.149

[4] Y. Sugiyama, T. Homae, K. Wakabayashi, T. Matsumura, and Y. Nakayama, Numerical simulations of blast wave characteristics with a two-dimensional axisymmetric room model, Shock Waves,27 (2017) 615-622. https://doi.org/10.1007/s00193-016-0706-4

[5] T. Homae, K. Wakabayashi, T. Matsumura, and Y. Nakayama, Blast pressure distribution around a wall, Sci.Tech.Energetic Materials, 72 (2011) 155-160.

[6]Y. Sugiyama, T. Homae, K. Wakabayashi, T. Matsumura, and Y. Nakayama, Numerical simulations on the propagation of blast wave around a wall, Sci.Tech.Energetic Materials, 75 (2014) 162-168.

Explosion Shock Waves and High Strain Rate Phenomena

Materials Research Proceedings **13** (2019) 57-62

Materials Research Forum LLC

https://doi.org/10.21741/9781644900338-10

Development of a Compact Container Protecting from Accidental Explosions of High Energy Materials

Teito Matsuishi[1,*], Kazuhito Fujiwara[1], Fumiko Kawashima[1], Haruka Oda[1]

[1]Kumamoto University, Chuo-Ku, Kurokami 2-39-1, Kumamoto, 860-8555 Japan

* 180d8476@st.kumamoto-u.ac.jp

Keywords: Ultra High Molecular Weight Polyethylene, Dyneema, Zylon, Compact Container, Suppressing Deformation

Abstract. There are many things in our surroundings that are at risk of explosion (e.g. lithium ion batteries, power modules, spray cans etc.), and there is a possibility of causing great damage to the surroundings due to explosive fragments. For safe operation of them, it is essential to establish a way to protect the surroundings from explosive fragments. In this study, the purpose is to develop a compact container protecting the surroundings from explosive fragments. Ultra-High Molecular Weight polyethylene (UHMW) which is excellent in strength against explosive and penetrationof fragments was used for container, and Dyneema string or Zylon string were wound around the container for suppressing deformation. In order to observe a deformation of containers due to explosion, containers were blown up by explosives, and taken by high speed camera. Experimental results showed that the supporting strings are available for suppressing deformation.

Introduction

It is expected that power modules will be popularized due to the influence of smart grid and soon, but when dielectric breakdown occurs when high voltage exceeding breakdown voltage is applied, explosion occurs due to high heat of chip of component parts have been confirmed. Besides, there are many things that have a risk of explosion around us, such as lithium ion batteries used in smart phones, headphones, etc. and spray cans used in paints, insecticides, etc. Because of the possibility of damage to the surrounding people and things due to the explosion and the fragments of the explosive component parts, it is essential to establish a method to protect the surroundings from explosion and fragments in order to operate safely.

As a method of protecting the surroundings from explosion and fragments, it is conceivable to develop a container that has strength against explosion and penetration. Strength against shock and high speed penetration of fragments is required for the material of the container. And also, in the case of lithium ion batteries or power modules, containers should be lightweight and compact to incorporate into the equipment in which they are used. As a material of container, UHMW which satisfies the above properties was used for container. Also, UHMW was reinforced by Dyneema string or Zylon string protecting from being broken due to an increase in strain. In order to investigate the influence of penetration of fragments, explosives were loaded in metal pipes and placed inside the container and blown up. High speed deformation due to explosion was observed by high speed camera. It was also investigated whether Dyneema and Zylon can be used as the material of the reinforcing layer. Experiments were conducted in Shock Wave Laboratory, Kumamoto University, Japan.

Explosion experiment of UHMW1

The deformation of UHMW due to explosion was observed. Sizes of UHMW used in the experiment and a schematic diagram of the experimental device are shown in Fig.1, and the device after assembly is shown in Fig.2. To observe the deformation of UHMW, took videos from the radial direction by using a high-speed video camera (PhantomV7.3). Table.1 shows experimental conditions. Detonating fuse (DF), which contains 9.5~11.5g/m pentaerythritol tetranitrate (PETN), was used as an explosive. In order to confirm whether Dyneema string (3.0 mm diameter) is available as a reinforcing material, container was exploded while changing the number of reinforcing layers. Also, to confirm the influence of the fragments, DF was installed in a brass pipe and exploded. Next, in order to observe the influence of the deformation due to the change of the explosive amount, explosives was changed to PETN, loaded in an explosive pipe and detonated with an electric detonator. Also, in order to observe the influence by the fragments, we made experiments while changing the material of the explosivepipe.

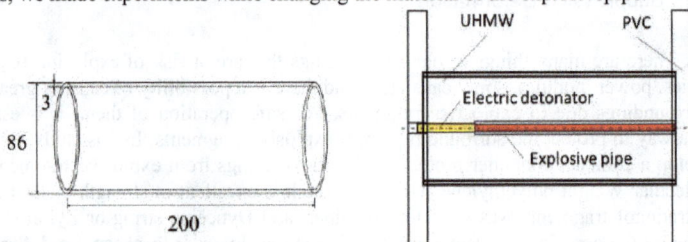

Fig.1 Sizes of UHMW, Schematic diagram of the experimental device

Table.1 Experimental conditions

	Explosive	Reinforcing layers	Brass pipe	Length of DF [mm]	Amount of explosive [g]	
No.1	DF	0	None	160	1.68	
No.2	DF	1	None	160	1.68	
No.3	DF	2	Used	160	1.68	
No.4	DF	2	None	190	1.995	
	Explosive	Reinforcing layers	Explosive pipe	Length of explosive part [mm]	Amount of explosive [g]	Inner diameter of cylinder [mm]
No.5	PETN	2	Straw	160	3.466	5.50
No.6	PETN	2	Aluminum	150	9.739	10.00
No.7	PETN	1	Copper	160	3.885	6.00
No.8	PETN	2	Copper	160	6.269	7.52

Fig.2 Experimental device after assembly

Explosion Shock Waves and High Strain Rate Phenomena

Materials Research Forum LLC

Materials Research Proceedings 13 (2019) 57-62

https://doi.org/10.21741/9781644900338-10

No.1　　　　　No.6　　　　　No.7

Fig3. Samples after exploded

Samples after exploded are shown in Fig.3. In the case of using DF, Only No.1 that was not reinforced by Dyneema was broken, and No.2, 3 and 4 that were reinforced showed no significant change. From these, it can be said that Dyneema is available for reinforcement material. In the case of using PETN, No.6 was broken. In No. 6, due to the increase in the amount of explosives, Dyneema was partially cut, and a part of UHMW expanded but wasn't broken. In the case of No.8, there were traces that the fragments penetrated at the end part, but wasn't broken. In No.5, no significant change was observed.

Explosion experiment of UHMW2
It was observed how the difference of material of explosive pipe affects the deformation of UHMW. ABS and copper pipes are used as a material of explosive pipe, conducted experiment while changing the amount of explosive. The material of the side plate was changed to SS400 in order to prevent the decrease in pressure due to the destruction. The experimental procedure is the same as in Chapter 2. A schematic diagram of the device is shown in Fig.4, and the device after assembly is shown in Fig.5. To observe the deformation of UHMW, took videos from the radial direction by Phantom V7.3. The experimental conditions are shown in Table.2. The reinforcement of Dyneema is 2 layers, and the length of explosive pipe is 160mm. In order to compare the influence of the differences in the material of the fragments, experiments using ABS for explosive pipe were conducted twice, and experiments using a copper pipe were conducted twice.

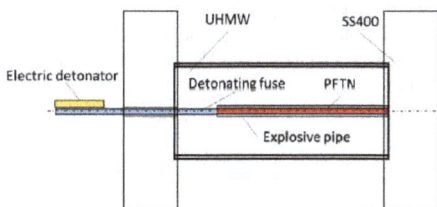

Fig.4 Schematic diagram of the device,　　　*Fig.5 Experimental device after assembly,*

Samples after exploded are shown in Fig.6. No.9 was not broken, but No.4, the amount of explosive was almost the same as No.9, was broken. No.10, which used the largest amount of explosives, was also broken. In the case of No.11 that used copper pipe as the material of explosive pipe and DF as explosive, it wasn't broken. Because the container was cracked from the part where the Dyneema was cut, it is thought that penetration of fragments greatly affects the destruction of the container.

Explosion Shock Waves and High Strain Rate Phenomena
Materials Research Proceedings 13 (2019) 57-62

Materials Research Forum LLC
https://doi.org/10.21741/9781644900338-10

Table.2 Experimental conditions

	Explosive pipe			Explosive	Amount of Explosive [g]
	Material	Inner dia [mm]	Thickness [mm]		
No.9	ABS	8	t2	PETN	6.9
No.10	ABS	11	t2	PETN	13.5
No.11	Copper	2.5	t1	DF	2.0
No.12	Copper	8	t1	PETN	7.0

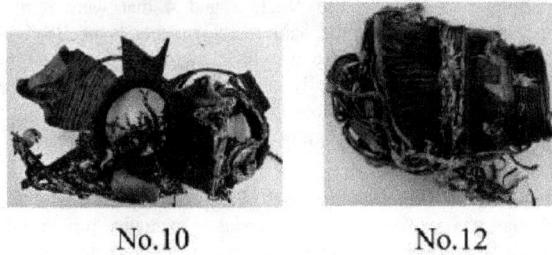

No.10   No.12

Fig.6 Samples after exploded

Study on reinforcement material

The influence of difference of reinforcement materials on the deformation of UHMW was compared. Besides Dyneema which has excellent tensile strength, ZylonX and Zylonknot, which have higher tensile strength than Dyneema, were used. The diameter of the reinforcement string is 3 mm for Dyneema, 1.9 mm for ZylonX, and 2.6 mm for Zylonknot. The size of UHMW used in the experiment and the schematic of the experimental device are shown in Fig.7. Copper was used for the material of explosive pipe, and the influence of fragments was observed. The experimental procedure is the same as in Chapter 2. To observe the deformation of UHMW, took videos from the radial direction by Phantom V 7.3. The experimental conditions are shown in Table.3, and physical properties of the reinforcing string material are shown in the Table.4, [1], [2]. The reinforcement for all samples is one layer.

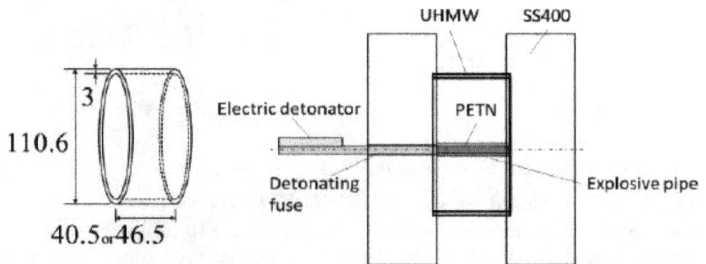

Fig.7 Sizes of UHMW, Schematic diagram of the experimental device

Table.3 Experimental conditions

	UHMW size [mm]	Reinforcing string		Copper pipe size [mm]	Amount of Explosive [g]
		Material	Diameter [mm]		
No.13	Inner dia 104.6 Outer dia 110.6 width 40.5	Dyneema	3	Inner dia 6 Thickness 1 width 30.5	0.736
No.14		ZylonX	1.9		0.734
No.15	Inner dia 104.6 Outer dia 110.6 width 46.5	Zylonknot	2.6	Inner dia 6 Thickness 1 width 36.5	0.860
No.16		Dyneema	3		0.848
No.17		Zylonknot	2.6	Inner dia 7.52 Thickness 1 width 36.5	1.397
No.18		Dyneema	3		1.408

Table.4 Physical properties of the reinforcing string material

	Tensile strength [GPa]	Tensile modulus [GPa]	Bulk modulus [GPa]
Dyneema	2.5	123	68.3
ZylonX	4.2	180	100
Zylonknot	4.0	120	66.7

Samples after exploded are shown in Fig.10. No.8 reinforced with ZylonX was broken, but UHMW was not broken in No.5, No.7 and No.8 which were exploded at almost the same amount of explosive, and reinforcing layer was cut at the part where the copper fragments was penetrated. No.17 and No.18 which used a large amount of explosives were also broken.

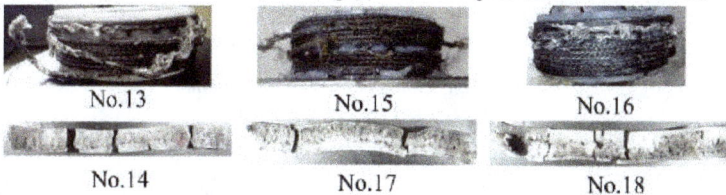

No.13   No.15   No.16

No.14   No.17   No.18

Fig.8 Samples after exploded

Influence of metal fragments on deformation
The influence of the amount and size of metal fragments on the deformation of UHMW was investigated. UHMW sizes and experimental procedures are the same as in Chapter 2. A schematic diagram of the experimental device is shown in Fig.9. The length of explosive pipe is 194 mm to make the fragments collide with UHMW uniformly. The explosive pipe was made of two layers, copper and ABS. The amount and size of fragments were adjusted by the diameter of the copper pipe, and the amount of explosives was adjusted by the diameter of the ABS pipe. The experimental conditions are shown in Table.5. Dyneema was used as the material of the reinforcing layer. The reinforcement is two layers.

A graph measuring the time change of the radius of the container due to the explosion is shown in Fig.10. Measured the central part of the container every 16µs, which is the frame interval of a high-speed camera. All samples were not broken, and containers were repeatedly expanding and contracting. As shown in Fig.7, No.17 which used a large copper pipe repeats expansion and contraction with large displacement. It is thought that larger fragments are generated if a large copper pipe is used, and causes a large deformation of the container.

Explosion Shock Waves and High Strain Rate Phenomena   Materials Research Forum LLC
Materials Research Proceedings **13** (2019) 57-62   https://doi.org/10.21741/9781644900338-10

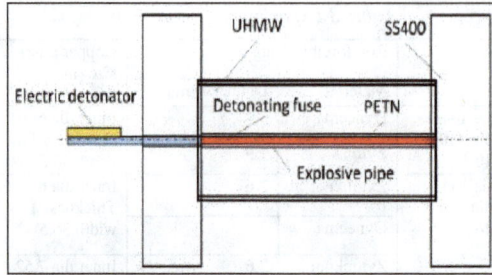

Fig.9 Schematic diagram of the experimental device

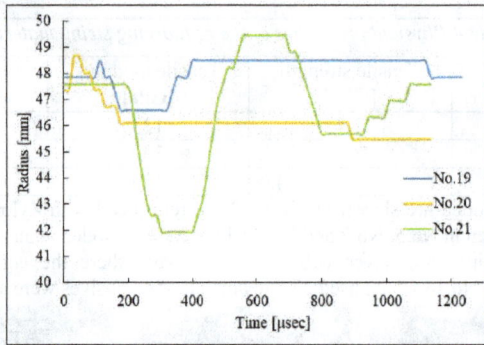

Fig.10 Time change of radius of container

Table.5 Experimental condition

| | Explosive pipe ABS | | Explosive pipe Copper | | Amount of explosive [g] |
	Inner dia [mm]	Thickness [mm]	Inner dia [mm]	Thickness [mm]	
No.19	4	t1	6	t1	2.156
No.20	6	t0.7	7.52	t1	4.354
No.21	6	t1	15	t1.5	4.217

Conclusions

It was confirmed that it is available as enforcing material for Zylon and Dyneema. It was found that the penetration of fragments has a great influence on the deformation of the container. It was found that the rigidity of the reinforcing layer material is related to the destruction of the container.

References

[1] TOYOBO.(n.d.). Product information of IZANAS, Rerived March 16, 2019,from http://www.toyobo.co.jp/seihin/dn/izanas/iz_product.html

[2] TOYOBO.(n.d.). Product information of ZYLON, Rerived March 16, 2019, from http://www.toyobo.co.jp/seihin/kc/pbo/zylon_features.html

Explosion Shock Waves and High Strain Rate Phenomena
Materials Research Proceedings **13** (2019) 63-67

Materials Research Forum LLC
https://doi.org/10.21741/9781644900338-11

Experimental and Theoretical Study of Fragment Safety Distance of Fragmenting Munitions

H.N. Behera[1, 2], Sarbjit Singh[2], Pal Dinesh Kumar[2], Asha Gupta[1]

[1]Panjab Engineering College (Deemed to be University), Sector 10, Chandigarh, India

[2]Terminal Ballistics Research Laboratory (TBRL-DRDO), Sector-30, Chandigarh 160030, India

hullash334@gmail.com, sarb59@gmail.com, ashagoel30@yahoo.co.in

Keywords: Fragment Safety Distance, Controlled Fragmentation, Fragment velocity, High Strain Rate

Abstract. The fragment safety distance is an important requirement for test and evaluation of the munition stores in the field trials. It determines the area to be cleared or evacuated before conduct of any trial activity. In this paper, theoretical and experimental work is carried out for establishing the explosive parameters and its interaction with the metallic casing. High explosives are used for controlled fragmentation to generate specific–size-and-weight fragments with lower velocity. Empirical relationship based on high strain rate and Gurney energy criteria were applied and optimized. Two prototypes having two different type explosive filling were fabricated to generate the fragment data. This enables to determine the safety distance useful for conducting trials in small ranges with required safety. The experimental data reveals that 90% fragments of a definite shape and size have been generated. The recorded fragment velocity was of the order of 250 to 400 m/s. Based on these data, safety distance was calculated and found to be about 400 m. Experimentally, fragments were recovered and found up to 130m from the point of burst.

1. Introduction

The objective of reducing fragment safety distance is to use the smaller ranges for trial and performance evaluation of fragmenting munitions. Fragment safety distance is a function of fragment mass, fragment velocity and fragments' aerodynamic behavior. If we can lower the fragment velocity and control fragment shape and size, then it is possible to reduce the fragment safety distance significantly. To do this, attempts are made to replace high explosive by high explosives for lower fragment velocity and controlled fragmentation technique [1-2] for definite shape fragment with known ballistic behaviors.

2. Theoretical Work

2.1. Particle Velocity and Maximum Pressure of the Shock Wave

The reduced pressure of the explosive filling inside the steel cylindrical warhead is calculated using Hugoniot values of various explosives, paraffin/High Energy Substitute (HES) and steel materials. The values [3] are given in Figure 1. The following equations are used for particle velocity and maximum pressure acting on the steel casing.

$$P_{CJ} = \rho_0 u_{JC} D \qquad \text{---(1)}$$

$$P = \rho_0 C_0 u + \rho_0 s u^2 \qquad \text{---(2)}$$

$$P = (2.412 \mathrm{P}_{CJ}) - (1.7315\, P_{CJ} / u_{CJ}) u + (0.3195\, P_{CJ} / u_{CJ}^2) u^2 \qquad \text{---(3)}$$

Where D=detonation velocity, P_{cj}= CJ Pressure, u = particle velocity, P=pressure

Explosion Shock Waves and High Strain Rate Phenomena   Materials Research Forum LLC
Materials Research Proceedings **13** (2019) 63-67   https://doi.org/10.21741/9781644900338-11

The equations 1- 3 are solved for particle velocity u and maximum pressure P of the shockwave.

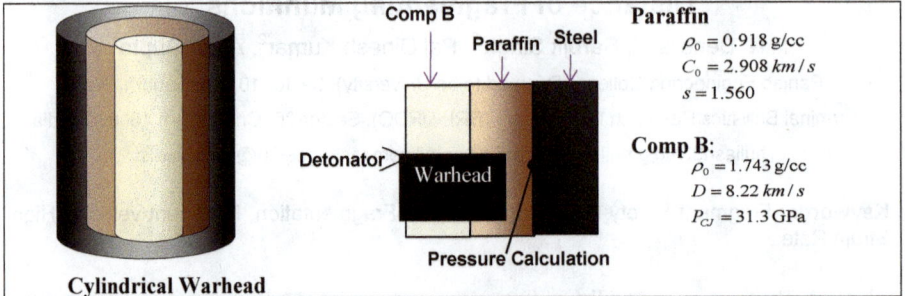

Figure 1 Pressure Calculation

2.2. Modified Gurney Formula for Initial Fragment Velocity

The initial fragment velocity is calculated using the Gurney equation [4]. But the Modified Gurney equation is also used to accommodate the ratio of warhead length to its internal diameter. The initial velocity of fragments resulting from a high-order detonation of a cylindrical warhead expressed as

$$V_0 = \sqrt{2E}\sqrt{\frac{C/M}{\left(1+\frac{0.5C}{M}\right)\left(1+\frac{0.5Di}{L}\right)}} \qquad \text{---(4)}$$

Where

V_0 = Initial velocity of fragments (m/s), $\sqrt{(2E)}$ = Gurney Constant (m/s)
C= Charge mass (g), M=casing mass (g), L=warhead length (mm), Di=internal diameter (mm)

2.3. Fragment Trajectory

The point mass trajectory model is used to compute the trajectory of a fragment with constant mass [5-6]. The fragment in air experiences the aerodynamic drag force and gravity. The motion of the fragment is described by the following differential equations:

$$\dot{V}_x = -\frac{\rho A_p C_d}{2m}VV_x$$

$$\dot{V}_y = -\frac{\rho A_p C_d}{2m}VV_y - g$$

$$\dot{V}_z = -\frac{\rho A_p C_d}{2m}VV_z \qquad \text{---(5)}$$

Where

m=fragment mass(g), V=magnitude of velocity vector(m/s),V_x,V_y,V_z=components of velocity of fragments,g=gravitational acceleration(m/s^2), ρ=air density(g/cc), Cd=drag coefficient of fragment,Ap=presented area of fragment(mm^2).

Explosion Shock Waves and High Strain Rate Phenomena   Materials Research Forum LLC
Materials Research Proceedings **13** (2019) 63-67   https://doi.org/10.21741/9781644900338-11

3. Experimental Work

Based on the theoratical calculations, two prototypes of 3 kg class controlled fragmentation warhead having two different type explosive filling were fabricated to generate the fragment data. Both prototypes were subject to static trial under similar conditions. Velocity measurement system was used to record the fragment velocity and strawboards for fragment recovery. The survey method was used to trace the fragments up to 250m distance from the point of burst. The trial setup is shown in Figure 2.

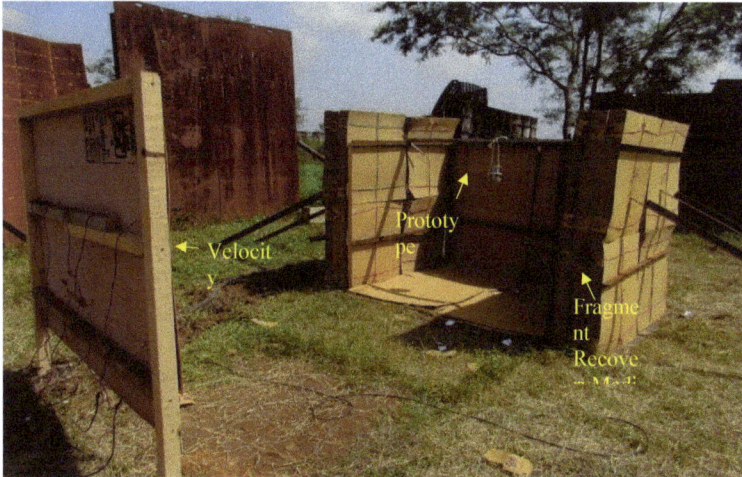

Figure 2 Trial Setup

4. Result & Analysis

Based on pressure and initial velocity of fragments, different ratios of diameters of high explosive substitute to that of high explosive were calculated.

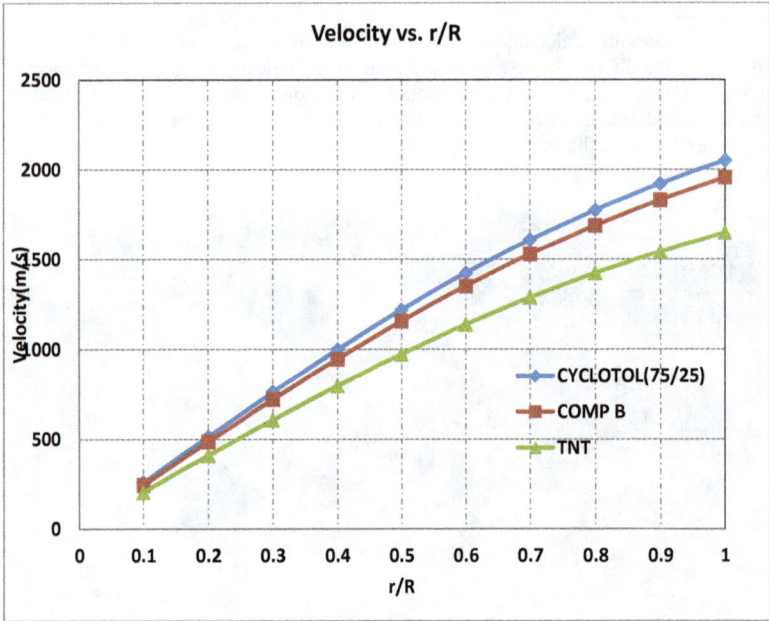

Figure 3. Fragment Velocity vs Filling Pattern

In Figure 3.The Curves are plotted using reduced pressure for different combinations of ratios of three explosive with HES paraffin. The curve of Fragment velocity vs different ratios of three explosive to HES are used to compute the optimum ratios for lower the fragment velocity. As an example, the composition(TNT:HES) gives the fragment velocity up to 600 m/s.

Based on these calculations, prototypes was fabricated and tested. Experimentally generated fragments are shown in Figure 4.

Figure 4. 3-kg class Prototype Warhead and Fragments

Using the fragment trajectory equations, fragments with highest velocity and diferent orientations were used to compute the fragment safety distances. The computed safety distance is about 400 m. The survey method was used to find the trace of fragments on the ground. Experimentally, it was observed that the fragments travelled up to 130m.

Conclusions

Two prototypes having two different type explosive filling were fabricated to generate the fragment data. This enables to determine the safety distance useful for conducting trials in small ranges with required safety. The experimental data reveals that 90% fragments of a definite shape and size have been generated. The fragment velocity is recorded in the order 250 to 400 m/s. Based on these data, safety distance was computed. The calculation shows that maximum safety distance is about 400 m. The experimetal observation shows that fragments travelled upto 130m.

Acknowledgements

The authors would like to gratefully thank Dr. Manjit Singh, DS and Director TBRL for his support and encouragement for this research work. The authors are also very thankful to all the supporting trial team members. This research did not receive any specific grant from any funding agencies in the public, commercial, or not-for-profit sectors.

References

[1] W Arnold, "Controlled Fragmentation", Shock Compression Matter,2001

[2] D Villano and F Galliccia "Innovative Technologies for Controlled Fragmentation Warheads",27th International Symposium on Ballistics,Freiburg,Germany,2013. https://doi.org/10.1115/1.4023341

[3] R. M. Lloyd, "Conventional Warhead System and Engineering Design", Progress in Astronautics and Aeronautics, Volume 179.1991

[4] R. W. Gurney, "The Initial Velocities of Fragments from Bombs, Shells and Grenades" BRL Report No 405, 1943. https://doi.org/10.1115/1.4023341

[5] Robert L McCoy, "Modern Exterior Ballistics – the Launch and Flight Dynamics of Symmetric Projectiles", 1999

[6] F.McCleskey., "Drag Co-efficients for Irregular Fragments,", NSWC TR 87-89,Naval Surface Warfare Centre, Dahlgreen, Virginia, February 1998

Explosion Shock Waves and High Strain Rate Phenomena
Materials Research Proceedings **13** (2019) 69-73

Materials Research Forum LLC
https://doi.org/10.21741/9781644900338-12

Joining of Dissimilar Metals Using Low Pressure Difference

Ayumu TAKAO[1, a *], Ryuichi TOMOSHIGE[2, b] and Akio KIRA[3, c]

[1]Graduate school of Engineering, SOJO University, 4-22-1 Ikeda, Nishi-ku, Kumamoto 860-0082, Japan

[2]Department of Nanoscience, SOJO University,4-22-1 Ikeda, Nishi-ku, Kumamoto 860-0082, Japan

[3]Department of Mechanical Engineering, SOJO University,4-22-1 Ikeda, Nishi-ku, Kumamoto 860-0082, Japan

[a]g1813m06@m.sojo-u.ac.jp , [b]tomosige@nono.sojo-u.ac.jp, [c]akira@mec.sojo-u.ac.jp

Keywords: Metal Bonding, Oblique Collision, Projectile Accelerator, Metal Jet

Abstract. In explosive welding, the velocity of flyer plate requisite for joining of two different kinds of metallic sheet is several hundred meters per second. We thought that the velocity would be accomplished easily without explosives. A lightweight projectile, which receives higher pressure on the rear side than front side, goes forward and is accelerated to extremely high velocity, even if the pressure difference is small. Joining should be achieved, when a thin metal sheet attached on the front of the projectile collides with another metal plate fixed on an oblique block. Oblique collisions between several kinds of metal were examined. Examinations of the joint interfaces of this resultant by both scanning electron and optical microscopes find no opening. Detachment at the joint interface did not occur, when tensile forces were applied. Therefore, we regard that the joint interface has sufficient strength.

Introduction

High-energy-rate processing has many excellent features that differ from static processing. For example, explosive welding, which is one of the methods for producing cladding materials, is applied to combinations in dissimilar metals and non-metals that are difficult to bond in diffusion bonding. Metal processing and material synthesis have been carried out with shock waves generated by explosives [1, 2]. The method that does not use explosives was originated, as experiments using explosives require qualifications to handle them and the cost of the experiment is high. The general methods of joining metals are mechanical bonding, metallurgical bonding, and chemical bonding. Each has advantages and disadvantages, and it is necessary to select a bonding method suitable for the material and bonding conditions to join metals efficiently. Explosive welding has the best features among these bonding methods. A simple projectile accelerator using a difference in air pressure has been produced. The equivalent qualifications to explosive welding will be succeeded, if the device were used. When this device is applied to sheet metal forming, good results than expected was obtained. Then I tried joining of dissimilar metals.

Experiment

In explosive welding, the flyer plate is arranged in parallel with an appropriate distance from the parent plate, and one end of the explosive placed on the flyer plate is detonated. In the proposed method, the flyer plate is accelerated by the air pressure difference substitute for the explosive. Since a high pressure difference is required to obtain a large acceleration, a vacuum collision chamber and a high-pressure chamber are made. Figure 1 is a photograph of the overall view of the projectile accelerator. A metal plate attached to the flyer plate is accelerated by the pressure difference, and it collides with another metal plate at high speed obliquely. In order to achieve this, as the condition called weldability window [3] must be satisfied, it is necessary to adjust the collision velocity V_c and the collision angle β appropriately according to the combination between the material of flyer plate and that of parent plate.

Fig.1 Overall view of the projectile accelerator.

Results

Figures 2-5 indicate the results of observation of the bonding interface of the joined metal plates obtained in this experiment with an optical microscope and an electron microscope. Figure 2 shows the results in the case of joining copper to another copper. Figures 3-5 are in the cases of joining copper to aluminum, aluminum to stainless steel, and aluminum to cast iron, respectively. In all case, no clearance or crack was observed, and they were joined well.

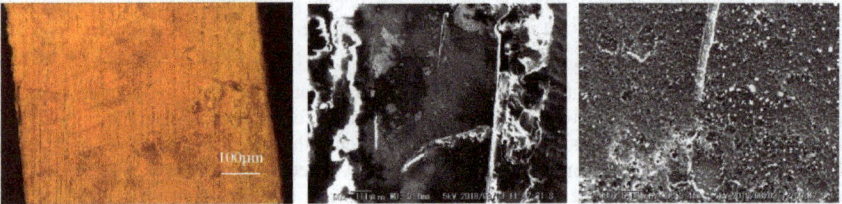

(a) Optical.   (b) Low magnification SEM.   (c) High magnification SEM.

Fig.2 Optical and SEM micrographs of bonding interface in the case of Cu and Cu.

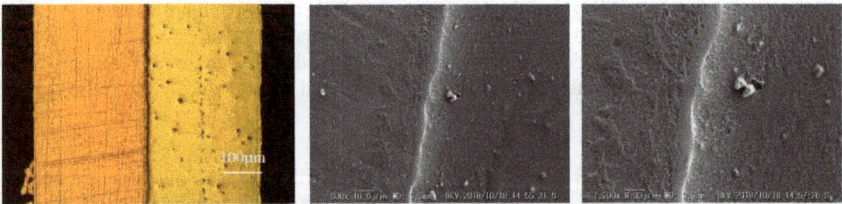

(a) Optical.   (b) Low magnification SEM.   (c) High magnification SEM.

Fig.3 Optical and SEM micrographs of bonding interface in the case of Cu and Al.

(a) Optical.   (b) Low magnification SEM.   (c) High magnification SEM.

Fig.4 Optical and SEM micrographs of bonding interface in the case of Al and SUS304.

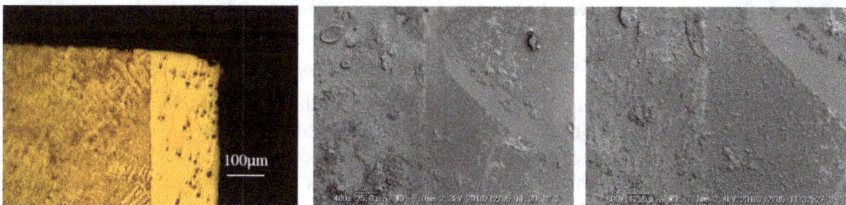

(a) Optical.   (b) Low magnification SEM.   (c) High magnification SEM.

Fig.5 Optical and SEM micrographs of bonding interface in the case of Al and cast iron.

Discussions

As the area between copper and aluminum in Fig.3(a) has different quality from two materials, intermetallic compounds may be formed by collision. Energy dispersive X-ray spectrometry (EDS) and X-ray diffraction analyzed (XRD) were operated. The central horizontal part with intermediate color in Fig.6(a) is the analysis target. Figures 6(b) and (C) indicate the molecular distribution map of copper and aluminum. The distribution of copper overlaps that of aluminum. Though the analysis of the crystal structure near the bonding interface was carried out by XRD, no peak for intermetallic compounds was observed.

(a) SEM of Al and Cu.   (b) Distribution of Al.   (c) Distribution of Cu.

Fig.6 Molecular distributions of boundary layer of Cu and Al observed with EDS.

In order to observe the bonding interface in detail, the surface was corroded. A photomicrograph of the corroded interface between copper and aluminum is shown in Fig.7(a). Figures 7(b) and (c) are in the case of copper and copper. In each case, as the procedure of corrosion was not appropriate, the structure could not be clearly observed.

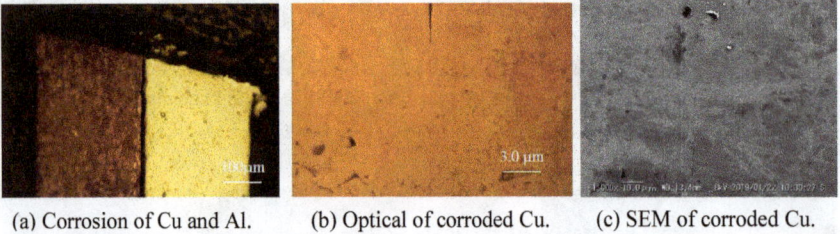

(a) Corrosion of Cu and Al.   (b) Optical of corroded Cu.   (c) SEM of corroded Cu.

Fig.7 Optical microscope and scanning electron microscope observation after corrosion.

A tensile test was carried out to check the bonding strength. Since a bonding material that can pull the bonding surface vertically cannot be manufactured, two plates with same size and material are aligned and another metal plate is bonded to the center of that. Figure 8(a) shows the test piece mounted on the tensile test instrument and (b) is the test piece after fractured. The stress σ is nominal stress that is obtained by dividing the maximum tensile load F at fracture point by the original cross-sectional area A, that is $\sigma = F/A = 59$ MPa. The result is slightly smaller than the tensile strength of aluminum 78 MPa. A slight gap between the two plates may cause a step on the flyer plate, which may affect the test results, and the bonding strength seems to be sufficient because it did not separate at the bonding interface as shown in Fig.8(b). Figure 8(c) is a curve plotted with nominal stress-nominal strain.

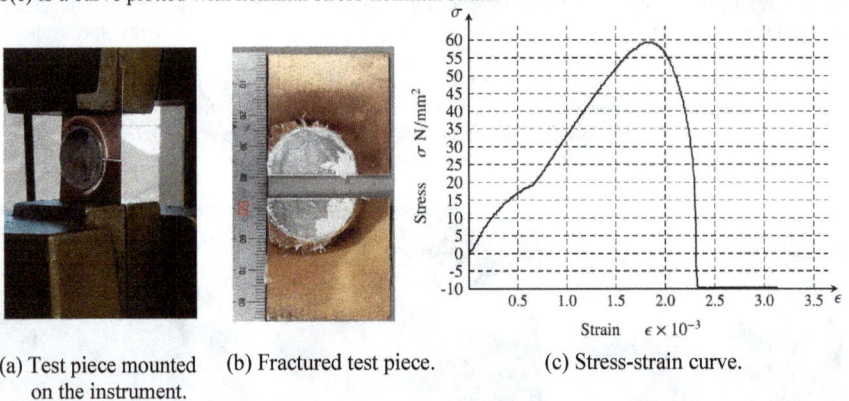

(a) Test piece mounted   (b) Fractured test piece.   (c) Stress-strain curve.
    on the instrument.

Fig.8 Tensile test.

Summary

In this study, we tried to join dissimilar metals using low-pressure difference.

- As in the observation at bonding interface by microscope, no clearance or crack was existed, bonding were success.

Explosion Shock Waves and High Strain Rate Phenomena   Materials Research Forum LLC
Materials Research Proceedings 13 (2019) 69-73   https://doi.org/10.21741/9781644900338-12

- Though in the observation by microscope, there was a different part in texture from original material, no peak for intermetallic compounds was observed in the analysis by XRD.
- The bonding strength is sufficient because it did not separate at the bonding interface in the tensile test.

References

[1] A. Kira, R. Tomoshige, K. Hokamoto and M. Fujita, Phase Transformation of Powdered Material by Using Metal Jet, Materials Science Forum, Vols.706-709, pp.741-744 (2012). https://doi.org/10.4028/www.scientific.net/msf.706-709.741

[2] A. Kira, K. Hokamoto, Y. Ujimoto, S. Kai and M. Fujita, Observation of Metal Jet in Extremely High Impulsive Pressure Generator, International Journal of Modern Physics B, Vol. 22, No. 9, 10 and 11, pp.1653-1658 (2008). https://doi.org/10.1142/s0217979208047213

[3] M. M. Hoseini Athar, B. Tolaminejad, Weldability window and the effect of interface morphology, Materials and Design, Vol.86, pp.516-525 (2015). https://doi.org/10.1016/j.matdes.2015.07.114

Explosion Shock Waves and High Strain Rate Phenomena
Materials Research Proceedings **13** (2019) 74-78

Materials Research Forum LLC
https://doi.org/10.21741/9781644900338-13

Observation for the High-Speed Oblique Collision of Metals

Akihisa Mori[1,a] *, Shigeru Tanaka[2,b] and Kazuyuki Hokamoto[3,c]

[1] Faculty of Engineering, Sojo University, Ikeda 4-22-1 Kumamoto, 860-0082 Japan

[2] Institute of Pulsed Power Science, Kumamoto University, 2-39-1 Kurokami, Chuo-ku, Kumamoto, 860-8555 Japan

[a]makihisa@mec.sojo-u.ac.jp, [b] tanaka@mech.kumamoto-u.ac.jp,

[c]hokamoto@mech.kumamoto-u.ac.jp

Keywords: Oblique Collision, Explosive Welding, Interfacial Waves, Metal Jet, Numerical Simulation

Abstract. In explosive welding, it is known well that the collision angle and collision velocity are the important parameters to achieve good welding. In addition, generations of a metal jet and the interfacial waves are important for the explosive welding conditions. To know the parameters and the collision conditions, the optical observation and the numerical simulation for the oblique collision using a powder gun were done by the authors. A metal jet was observed clearly by using a powder gun and wavy interface was generated without the intermetallic layer for the reactive materials by controlling the welding conditions. In this investigation, the results of the optical observations and the numerical analysis for similar and dissimilar material combinations were reported.

Introduction

Explosive welding technique is known well as the welding method to weld strongly for the two metal plates of similar and/or dissimilar material combinations. In explosive welding technique, a metal flyer plate is accelerated by the detonation of explosive and is collided to another metal plate (base plate) with a certain angle at high velocity. A good welding is achieved with generating the interfacial waves in the welded interface and the metal jet at the collision point when the velocity and the angle collided are within the suitable range [1, 2]. Therefore, to achieve the optimal welding conditions for the difficult-to-weld materials, it is necessary to know the parameters and the collision phenomena, such as the metal jet generations and the interfacial waves. The mechanism of interfacial waves and the metal jet generation have been studied theoretically and/or numerically by many researchers [3-5]. Onzawa et al. [6] reported about the characteristics of metal jet generated by the collision of similar and dissimilar metals set on parallel and angular arrangement using a high-speed streak camera. The observation for the metal jet generation is difficult by the optical observation system because the detonation gas spreads out rapidly with the high velocity which is faster than the flying velocity of metal. From the weldability window proposed by Wittman [7] and Deribas [8], claddings same as explosive welding can be obtained when a metal plate collides obliquely at high velocity. To know the inclined collision, same as the phenomena of explosive welding, a powder gun was applied to observe the high-speed oblique collision, which is same as the phenomena of explosive welding, without the influence of detonation gas. And the numerical simulation using SPH solver in ANSYS AUTODYN software was used to understand the material behavior in the high-speed oblique collision, comparing with the experimental results.

Explosion Shock Waves and High Strain Rate Phenomena   Materials Research Forum LLC
Materials Research Proceedings **13** (2019) 74-78   https://doi.org/10.21741/9781644900338-13

Experimental Procedure

Experimental setup to observe the high-speed oblique collision is shown in Figure 1. A powder gun set on Institute of Pulsed Power Science in Kumamoto University was used to accelerate the metal plate. A pure copper and magnesium alloy AZ31, which diameter were 32 mm and thickness was 3 mm or 5 mm, were applied as the flyer and target plate. The flyer plate was combined the sabot made by Ultra high molecular weight polyethylene (UHPE) as the projectile. The projectile was set in the barrel of the powder chamber side. The copper weight-control plate was placed behind the flyer plate to control the flying velocity of projectile. The target plate put into a PMMA target holder was arranged on the target stand with an inclined angle (θ = 7, 10, 15, 20) in the target chamber. For the optical observation of the oblique collision, High-speed video camera (HPV-1, Shimadzu corp., capable of recording up to 1 million fps) was placed at the side of the target chamber and was located in the opposite side of the light across the target chamber. Smokeless and the black gunpowder were set in the powder chamber. After the target chamber was in a vacuum, the black gunpowder was ignited.

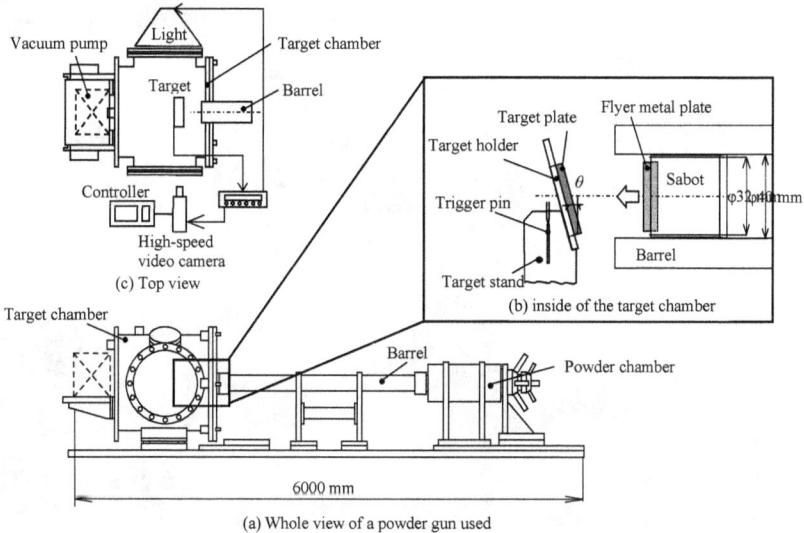

Fig. 1 Schematics of experimental setup used

Numerical Analysis

Explicit dynamics software ANSYS AUTODYN was used and the 2-dimensional planer symmetry was applied to know the detail of the oblique collision at high velocity numerically. A target plate and a projectile were modeled by two solvers, which were the Smoothed Particle Hydrodynamics (SPH) solver and the Lagrangian solver. The 60% thickness part on the collision side of the metal plate was modeled by the SPH solver, and the remaining 40% was applied by the Lagrangian. The particle size of SPH solver and the mesh size of Lagragian solver were fixed at 0.05 mm and 0.03 mm when the thickness of metal plates was 5 mm and 3 mm respectively. The Mie-Grüneisen form shock equation of stat and the Johnson-Cook strength model were applied for each material. The material parameters for each equation are referred from the reports [9, 10].

Materials Research Forum LLC
https://doi.org/10.21741/9781644900338-13

Results and discussion

Examples of framing photographs after the starting of collision were shown in Figure 2. The metal jets generation for the similar and dissimilar material combinations could be observed clearly in the oblique collision, which was the same condition in the explosive welding process, when a powder gun was used. The metal jets were generally dark and not bright when the copper and copper oblique collision. In the case of using a magnesium alloy regardless of the similar / dissimilar combination, bright metal jets were observed shown in Figure 2 (b) and (c). In addition, it was confirmed that two types of metal jet, the central jet and the surrounding jet, were generated when the installation angle of the plate was 15 degrees or more, as shown in Figure 2 (a) and (c).

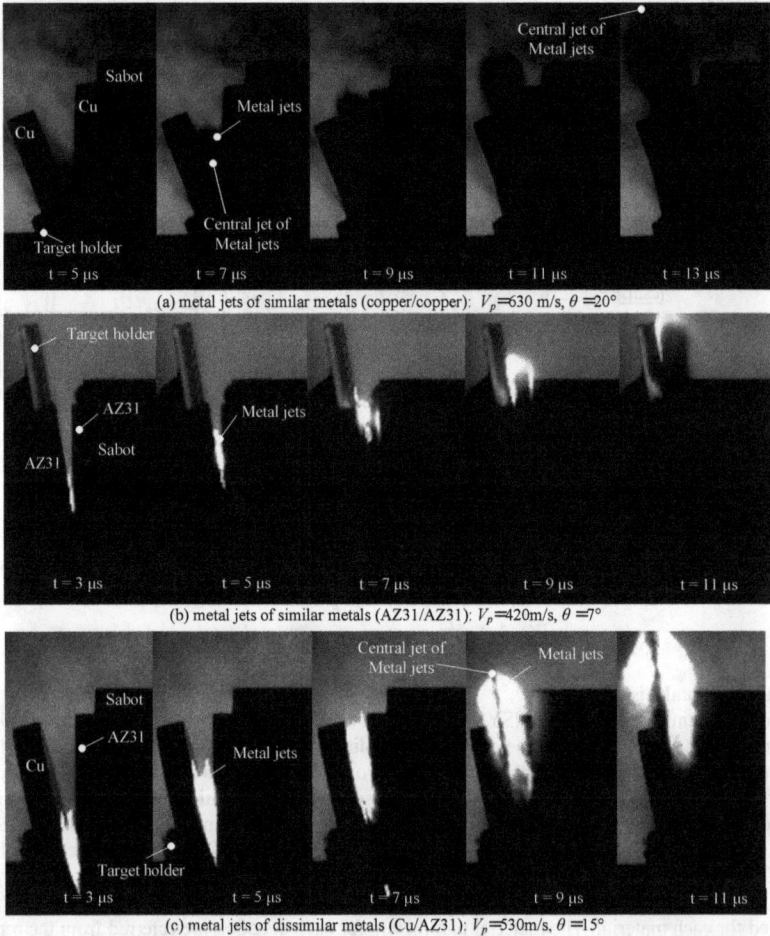

(a) metal jets of similar metals (copper/copper): V_p=630 m/s, θ =20°

(b) metal jets of similar metals (AZ31/AZ31): V_p=420m/s, θ =7°

(c) metal jets of dissimilar metals (Cu/AZ31): V_p=530m/s, θ =15°

Fig. 2 Metal jet generations in the oblique collsino of similar and dissimilar material combition

Explosion Shock Waves and High Strain Rate Phenomena
Materials Research Proceedings 13 (2019) 74-78

Materials Research Forum LLC
https://doi.org/10.21741/9781644900338-13

From the formula of the relation for the collision velocity, the collision point velocity and the collision angle [1], the collision point velocity was 1810, 3440 and 1780 m/s in the condition of Figure 2(a), (b) (c) respectively. And the moving velocity of the top of metal jets obtained from these photographs was 3540, 5350 and 3560 m/s, which was about twice the collision point velocity.

Numerical results for the oblique collision of Cu/Cu, AZ31/AZ31 and Cu/AZ31 using ANSYS AUTODYN-2D were shown in Figure 3. The particle velocity constituting the metal jets agreed with the experimental results. In the case of Figure 3(a), the large-size interfacial waves which height was about 1 mm were formed from the center to the end of metals numerically although the small interfacial waves were formed around the starting of the collision. On the other hand, in the case of magnesium alloys, even the similar material combination, small interfacial waves were simulated from the starting of the collision to end. In the temperature contour of the numerical results, the metal jets were increased at 2000 ~ 3000 K in the case of copper and copper metals and increase over 5000 K in the magnesium alloy. Since the melting point of the magnesium alloy is 650 °C and the boiling point is 1090 °C, it was thought that magnesium jets were vaporized instantaneously, considering only the temperature state. However, it was unclear what the phase of metal jets specifically was because of the high-pressure condition at the time of collision. A slight difference in the state of temperature and pressure has been speculated that two types of jet could be observed in experimental results. From the numerical analysis, the metal jets were composed only of magnesium alloy in the case of dissimilar oblique collision. It was possible that the jets were composed mainly of the low-density material.

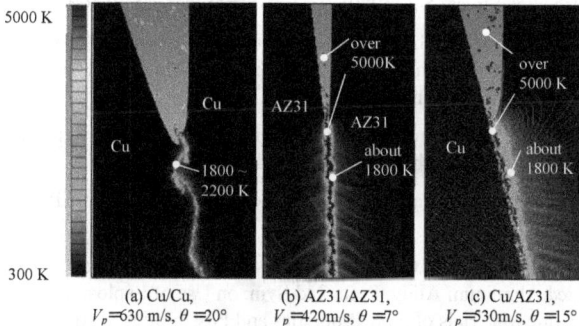

(a) Cu/Cu,
V_p=630 m/s, θ =20°

(b) AZ31/AZ31,
V_p=420m/s, θ =7°

(c) Cu/AZ31,
V_p=530m/s, θ =15°

Fig.3 Temperature contour of the collided interface obtained from the numerical simulation

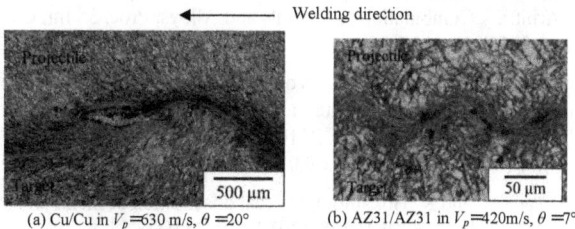

(a) Cu/Cu in V_p=630 m/s, θ =20°

(b) AZ31/AZ31 in V_p=420m/s, θ =7°

Fig.4 Welded interface of the recovered sample

Figure 4 shows a welding cross-section of recovered in experimental samples. In the similar metal combination, a good welding was achieved with forming the interfacial waves in the same as the explosive welding. However, a good welding was not achieved in the dissimilar metal combination. In the case of magnesium alloys, it could not be achieved to weld at high velocity or large collision angle. In the case of the Cu/Cu oblique collision, as shown in Figure 4(a), molten parts were observed in the welded interface. Therefore, it was thought that the separation or the destruction was occurred at the molten parts when a magnesium alloy was used.

Summary

The phenomena of high-speed oblique collision of the two metals were investigated in the experiments using the optical observation system and in the numerical simulation. In the optical observations metal jets were observed clearly and two types of metal jets were observed when the setup angle was 15 degrees and more. And it was confirmed that the top of metal jets was propagated at approximately twice of the collision point velocity. In the numerical results, it could be confirmed the behaviors and temperature conditions of the collided interface and the metal jets, which were difficult to measure by the experimental method.

References

[1] B. Crossland, Explosive Welding of Metals and its Application, Oxford University Press, 1982.

[2] M. A. Meyers, et al., Dynamic behavior of materials, John Wiley & Sons, 1994.

[3] C. Chemin, T. Qingming, Mechanism of Wave Formation at the Interface in Explosive Welding, Acta Mechanica Sinica, 5-2 (1989), 97-108. https://doi.org/10.1007/bf02489134

[4] S. Kakizaki, M. Watanabe and S. Kumai, Simulation and Experimental Analysis of Metal Jet Emission and Weld Interface Morphology in Impact Welding, Materials Trans., 52-5 (2011), 1003-1008. https://doi.org/10.2320/matertrans.l-mz201128

[5] G.R. Cowan, A.H. Holtman, Flow Configurations in Colliding Plates: Explosive Bonding, J. Appl. Phys., 34-4 (1963), 928-939. https://doi.org/10.1063/1.1729565

[6] T. Onzawa, Y. Ishii, Fundamental Studies on Explosive Welding, Trans. Japan Welding Society, 6-2 (1975), 98-104.

[7] R.H. Wittman, The Influence of Collision Parameters on the Strength and Microstructure of an Explosion Welded Aluminum Alloy, Proc. 2nd Sym. on Use of Explosive Energy in Manufacturing Metallic Materials of New Properties and Possibilities of Application thereof in the Chemical Industry, (1973), 153-168.

[8] A.A. Deribas, V.A. Simonov, and I.D. Zakcharenko, Investigation of Explosive Welding Parameters for Arbitrary Combinations of Metals and Alloys, Proc. 5th Int. Conf. on High Energy Rate Fabrication, (1975),4.1.1-4.1.24.

[9] Ulacia, I., Salisbury, C. P., Hurtado, I., Worswick, M. J., Tensile characterization and constitutive modeling of AZ31B magnesium alloy sheet over wide range of strain rates and temperatures, J. Mater. Process. Technol., 211(5), 830-839 (2011). https://doi.org/10.1016/j.jmatprotec.2010.09.010

[10] A. Mori, S. Tanaka, and K. Hokamoto, Optical observation of metal jet generated by high speed inclined collision, Proc. SPIE 10328, (2017), 103281Q. https://doi.org/10.1117/12.2270473

Explosion Shock Waves and High Strain Rate Phenomena
Materials Research Proceedings 13 (2019) 79-84

Materials Research Forum LLC
https://doi.org/10.21741/9781644900338-14

Effect of Added Molybdenum on Material Properties of Zr2SC MAX Phase Produced by Self-Propagating High Temperature Synthesis

RYUICHI Tomoshige[1, a *] KIYOHITO Ishida[2,b] and HITOSHI Inokawa[1, c]

[1] Sojo university, 22-1, Ikeda 4-chome, Nishi-ku, Kumamoto, 860-0082, Japan

[2] Tohoku University, 6-3, Aoba, Aramaki, Aoba-ku, Sendai, 980-8578, Japan

[a] tomosige@nano.sojo-u.ac.jp, [b] ishida@material.tohoku.ac.jp [c] inokawa@nano.sojo-u.ac.jp

Keywords: Combustion Synthesis, MAX Phase, Layered Structures, Carbosulfide

Abstract. Zr_2SC MAX phase with the layered structures was produced by self-propagating high temperature synthesis (SHS). Basic composition for MAX phase was determined in molar ratio of Zr:S:C=2:1:1. In addition, molybdenum of transition metal element was added according to the molar ratio of Zr:Mo:S:C=2-x:x:1:1 (x = 0 to 1.2) in order to attempt a formation of solid solution of MAX phase. SHS was initiated by using a metal heating coil. The synthesized materials were evaluated by XRD, Vickers hardness tests, SEM and TEM. XRD patterns of the synthesized MAX phases showed proof of formation of solid solution up to 20 at% of added molybdenum to zirconium. SEM observations revealed that the interlayer of monolithic Zr_2SC phase bonded strongly each other, and it looked like the structure in which it must be difficult to occur the interlayer exfoliation. On the other hand, molybdenum-added MAX phase had the interlayer structure at which it must be easy to exfoliate. TEM observations showed that the material was consisted of thin lamellas with about 10 nm thickness.

Introduction

In recent years, many researchers in inorganic materials science have been working on MAX phase materials in order to sleuth their unique properties. MAX phases are crystals of ternary nitrides or carbides with layered structures. Their general formula can be described as $M_{n+1}AX_n$ (where M is a transition metal, A is an A group (mostly IIIA and IVA) element, and X is N and/or C, and n = 1 to 3). Its features are typified by the combination of the properties of ceramics and metal materials, and it is remarkable in mechanical properties such as machining and deformability [1]. Also, Medkour et al. [2] and Pang et al. [3] reported in detail on electrical and thermal conductivities of MAX phase, respectively. Furthermore, Jovic et al. [4] reported the behavior of oxidation and corrosion of the material. In addition to the above-mentioned properties, it is well-known that treatments of MAX phase with an acid can elute selectively a layer A in MAX phase, which is formed from the group A element, and it has also been confirmed that an intercalation compound is formed from the elution-treated MAX phase. Thus, it is considered that these properties result from the crystal structure of MAX phase, especially, their layered structure. Graphite and MoS_2 which are used as solid lubricants, are also known as the materials having the layered structure. Their lubricating property is attributable to exfoliation at relatively weak bonding sites in the layered compound. Although the lubricating properties of MAX phase have not been clearly reported, it is speculated that the materials also has good lubricating property due to their layered crystal structure. On the other hand, it has been known to change properties according to the crystal structure, not only in the MAX phase but other materials. In particular, formation of a solid solution is known extensively as a method capable of artificially changing the crystal structure. Interstitial and substitutional solid solutions can be formed by adding an element different from the constituent elements. Naguib et al. introduced many types of solid solutions in the MAX phase, but they are mainly the cases of titanium-based MAX phase materials, and the aspect on formation of solid solution has not been clarified in other MAX phases [5]. Further, it is not readily to find the material properties in the case of MAX phase including sulfur as a constituent element because of their difficulty in synthesizing itself. By the way, it is well-known that self-propagating high temperature synthesis (SHS) is a method of producing spontaneously compounds with high temperature. One of the authors has also studied the synthesis of chromium sulfide using SHS. It was observed a phenomenon that a part of sulfur added as a raw material volatilized during synthesis under high temperature. However, the sulfides could be synthesized successfully without precisely controlled sintering method [6].

In this research, focusing on Zr-S-C MAX phase, it has been investigated whether synthesis of solid solutions of the carbosulfide is possible by SHS, and some of their material characteristics were evaluated.

Experimental Procedure

Commercially available powders of zirconium (Size:150 μm, Purity: 98%, Kojundo Chemical Laboratory), graphite (10 μm, 99.8%, Kojundo Chemical Laboratory), sulfur (75 μm, 99.5%, Kanto Chemical) and molybdenum (1 to 2 μm, 99.9%, Sigma-Aldrich) were used as raw materials, and were wet-mixed in ethanol for 1 hour in molar ratio of $Zr:Mo:C:S = (2-x):x:1:1$ ($x = 0$ to 1.2), followed by sufficient drying the mixtures. Here, molybdenum was selected as added solute element, which has the ion radius, Mo^{4+}, of 0.068nm similar to zirconium (Zr^{4+}: 0.087nm). It was also expected the addition of molybdenum with higher melting point than zirconium to suppress intensive reactions during synthesizing MAX phases. The powder mixtures were hydraulic pressed for molding pellets with 12mm in diameter and 10mm in height. The pellets were embedded into the steel container filled with sand, then SHS was carried out with a nichrome heating-coil settled at the top of the pellet. After applying voltage of 10V for initiation of SHS, the energetic chemical reaction was completed in a few seconds. The synthesized materials were investigated on microstructural observations by scanning (ERA-600, ELIONIX, Japan) and transmission electron microscopy (TITAN Themis, FEI, USA), phase identification

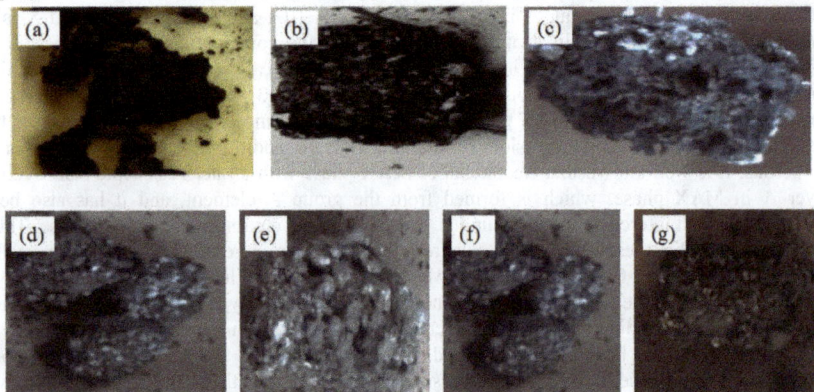

Fig.1 Outer views of obtained samples with a composition of Zr:Mo:S:C = 2-x: x:1:1 ; (a) x = 0, (b) 0.2, (c) 0.4, (d) 0.6, (e) 0.8, (f) 1.0 and (g) 1.2.

Materials Research Forum LLC
https://doi.org/10.21741/9781644900338-14

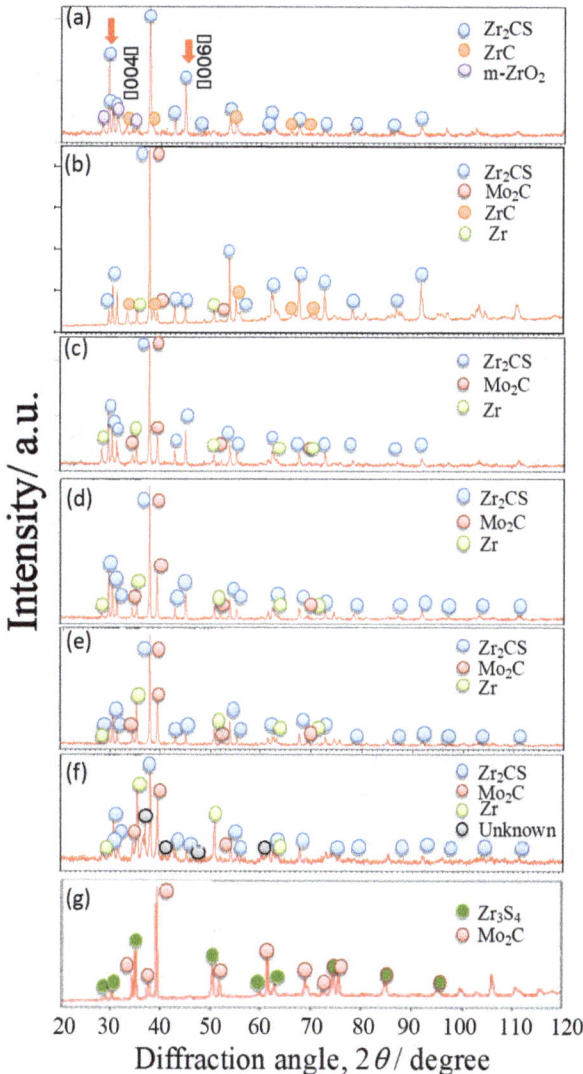

Fig.2 XRD patterns of combustion synthesized samples with a composition of Zr:Mo:S:C = 2-x: x:1:1 ; (a) x = 0, (b) 0.2, (c) 0.4, (d) 0.6, (e) 0.8, (f) 1.0 and (g) 1.2.

by X-ray diffraction (X'Pert Pro, Panalytical, the Netherlands), and Vickers hardness tests (HV-100, Mitsutoyo, Japan).

Results and discussion

Figure 1 shows outer views of the samples obtained by SHS. All of them were not in their original form due to their very energetic chemical reaction. Some of the specimens appeared to be melted. Comparatively, addition of molybdenum powder inhibited the reaction activity, and the specimens obtained by this synthesis became porous and brittle with increasing the molybdenum content.

Figure 2 shows the results of XRD experiments. In specimen (a) without molybdenum, Zr_2CS was formed with a small quantity of ZrC phase and trace of ZrO_2. Their diffraction peaks from the specific planes of (004) and (006) indicated very high intensity comparing with ICDD data. It is considered that the material had strong anisotropy. XRD patterns in the specimens (b) through (f) were similar to each other. Unlike in the case mentioned above, it was indicated that Zr_2SC MAX phase disappeared and only Zr_3S_4 and Mo_2C phases were formed dominantly in the diffraction pattern of specimen (g), which was the most molybdenum-added specimen in this study. This means Zr_2SC MAX phase cannot be maintained under the existence of excess molybdenum. The ratio of each product was quantified with peak intensity from the formed phases as shown in Figure 3. Vertical axis values indicated the ratio of diffraction peak height from (111) plane of ZrC, or (101) plane of Mo_2C to that from (004) plane of Zr_2SC. The results revealed that the amount of ZrC phases decreased and Mo_2C increased as adding molybdenum. This means that ZrC was formed as a bi-product of Zr_2SC phase when x=0 and 0.2. On the other hand, Mo_2C phase became a main phase in the range of x=0.4 to 1.2. It seems to be dominant that Zr and C reacted each other to form ZrC at the synthesis stage rather than that molybdenum extracted carbon from Zr_2SC phase synthesized by SHS in order to ZrC. Next, variation in lattice spacing of Zr_2SC phase is shown in Figure 4. Measurement of the lattice spacing was carried out on the base of diffraction peak from (006) plane of Zr_2SC. It was found that the lattice spacing varied remarkably between x=0 and 0.2 in the composition of $Zr:Mo:S:C=2-x:x:1:1$. However, the spacing hardly varied in the composition of more than x=0.4. Therefore, solid solution limit of molybdenum must be up to x=0.2, and it is considered the substitution cannot generate at the higher ratio of molybdenum.

Hardness was tested as a part of mechanical properties of Zr-Mo-S-C system MAX phase as shown in Fig.5. The test was performed under the condition of load of 4.9N and loading time of

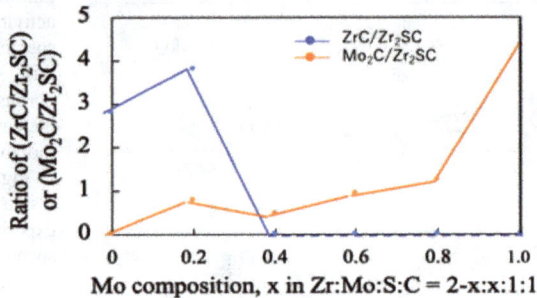

Fig.3 Variation in existence ratio of the formed phases

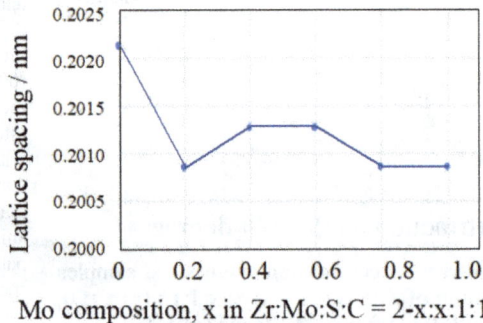

Fig.4 Variation in lattice spacing of (006) of Zr_2SC phase

Explosion Shock Waves and High Strain Rate Phenomena
Materials Research Proceedings **13** (2019) 79-84

Materials Research Forum LLC
https://doi.org/10.21741/9781644900338-14

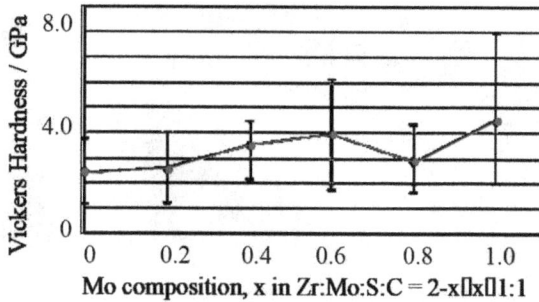

Fig.5 Results of the hardness test

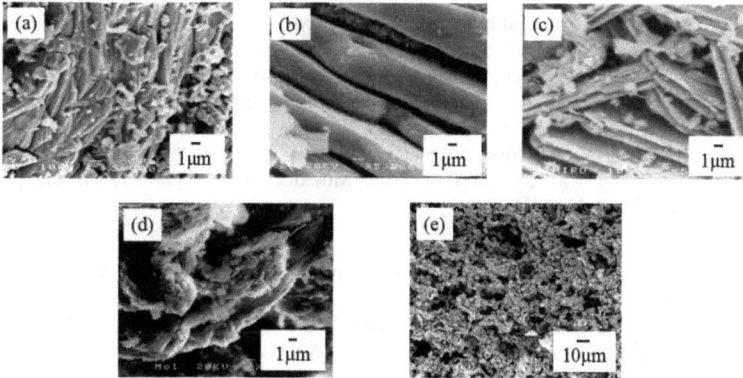

Fig.6 SEM observations of the obtained samples with a composition of Zr:Mo:S:C
= 2-x: x:1:1 ; (a) x = 0, (b) 0.2, (c) 0.4, (d) 0.6, (e) 0.8 and (f) 1.0.

30 sec. The hardness increased slightly with the addition of Mo. It was thought to attribute mostly to hardness of Mo_2C though there was some influence of the solid solution hardening.

Figure 6 shows SEM micrographs of various specimens. Figs.6 (a) to (c) indicated the layered structures. Especially, in Fig.6 (c), fine bi-layered structure was observed. The layered structure is preferable for a solid lubricant. Excess addition of the molybdenum, more than x=0.6 in this case, caused the specimen to be porous and rugged (Figs.6 (d, e)). TEM observations were also carried out for clarifying their nanostructure. Figure 7(a) and (b) revealed that Zr_2SC phase without molybdenum had the layered structures, which consisted of belt-like crystals in the order of 40 to 50 nm in width. In addition, they looked like to be in close contact each other. On the other hand, TEM image shown in Fig.7(c), which was taken from the sample with a composition of Zr:Mo:C:S=1.6:0.4:1:1, revealed that the sample had also the layered structure, but its layer thickness was around 10 nm. The characteristic layered structure could be seen in common between MAX phase in this study and MoS_2, which is well-known as typical solid lubricant. For example, thickness and flexibility enough to bend at a relatively steep angle are similar to each other [7,8]. Therefore, it is expected that the MAX phase material may become good solid lubricant.

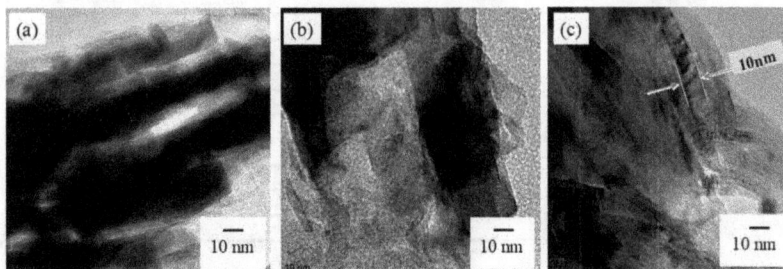

Fig.7 TEM observations of the obtained samples with a composition of (a,b)
Zr:Mo:S:C = 2:1:1, and (c) Zr:Mo:S:C= 1.6:0.4:1:1.

Conclusions

1. Zr_2SC could be produced readily by SHS, however with ZrC and Mo_2C of by-products.
2. It was speculated the solid solution could be formed up to 20% of molybdenum to zirconium.
3. It was considered that hardening of the composite was mainly attributed to the formation of Mo_2C, and that influence of formation of the solid solution on hardening was small.
4. TEM observations revealed that Mo-added MAX phase consisted of thin layers 10nm wide.
5. Because the material properties of Mo-added Zr_2SC MAX phase were different from those of monolithic Zr_2SC, it was concluded that the Mo-added one might become the promising material, especially in the solid lubricity.

References

[1] M. Radovic, M.W. Barsoum, MAX phases: Bridging the gap between metals and ceramics, Am. Ceram. Soc. Bull. 92(3) (2013) 20-27.

[2] Y.Medkour, A.Roumili, D.Maouche, L.Louail, 7- Electrical properties of MAX phases, in: M Low (Eds.), Advances in Science and Technology of Mn+1AXn Phases, Woodhead Publishing, Cambridge, 2012, pp. 159-175. https://doi.org/10.1533/9780857096012.159

[3] W.K. Pang, I.M. Low, Understanding and improving the thermal stability of layered ternary carbides in ceramic matrix composites, in: M Low (Eds.), Advances in Ceramic Matrix Composites, Woodhead Publishing, Cambridge, 2014, pp 340-368. https://doi.org/10.1533/9780857098825.2.340

[4] V. Jovic, M. Barsoum, Corrosion Behavior and Passive Film Characteristics Formed on Ti, Ti_3SiC_2, and Ti_4AlN_3 in H_2SO_4 and HCl, J. Electrochem. Soc. Vol.151(2) (2004) B71-B76. https://doi.org/10.1149/1.1637897

[5] M. Naguib, G. W. Bentzel, J. Shah, J. Halim, E. N. Caspi, J. Lu, L. Hultman, M. W. Barsoum, New Solid Solution MAX Phases: $(Ti_{0.5}, V_{0.5})_3AlC_2$, $(Nb_{0.5}, V_{0.5})_2AlC$, $(Nb_{0.5}, V_{0.5})_4AlC_3$ and $(Nb_{0.8}, Zr_{0.2})_2AlC$, Mater. Res. Lett. 2(4) (2014) 233-240. https://doi.org/10.1080/21663831.2014.932858

[6] R.Tomoshige, K. Niitsu, T.Sekiguchi, K.Oikawa, K.Ishida, Some Tribological Properties of SHS-Produced Chromium Sulfide, Int. J. self-propagating high temperature synthesis, 18 (4) (2009) 287-292. https://doi.org/10.3103/s1061386209040104

[7] V.Buck, Morphological properties of sputtered MoS_2 films, Wear. 91 (1983) 281–288. https://doi.org/10.1016/0043-1648(83)90073-x

[8] K.J.Wahl, I.L. Singer, Quantification of a lubricant transfer process that enhances the sliding life of a MoS_2 coating, Tribol. Lett. 1 (1995) 59–66. https://doi.org/10.1007/bf00157976

Explosion Shock Waves and High Strain Rate Phenomena
Materials Research Proceedings 13 (2019) 85-90

Materials Research Forum LLC
https://doi.org/10.21741/9781644900338-15

FTMP-Based Quantitative Evaluations for Dynamic Behavior of Dislocation Wall Structures

Shiro IHARA[1, a *], Tadashi HASEBE[1,b]

[1]1-1, Rokkodai-cho, Nada-ku, Kobe, Hyogo, Japan

[a]ihara.shiro@mail.mm4.scitec.kobe-u.ac.jp, [b]hasebe@mech.kobe-u.ac.jp

Keywords: Dislocation, Crystal Plasticity, Field Theory, Differential Geometry, Dislocation Dynamics

Abstract. Field theory of multiscale plasticity (FTMP) is applied to the quantitative evaluations of geometrically necessary boundaries (GNBs) of dislocations. Reproduced four representative GNBs via dislocation dynamics simulations are scrutinized via the duality diagram representation scheme and the Shannon entropy. Notable correlations are found both between the entropy and the incompatibility, and the mean entropy rate and the incompatibility rate. Furthermore, the mean entropy rate is linearly correlated also with the log of the incompatibility. Combined with a unified correlation for all the GNBs on the duality diagram, we demonstrate the effectiveness of the FTMP-based stability/instability criterion proposed in the previous study.

Introduction

Evolving dislocation substructures are one of the key features in achieving practically-feasible multiscale modeling of crystalline metallic materials. Geometrically necessary boundaries (GNBs) of dislocations[1,2] are classified as the typical wall structures among others that play pivotal roles not only for determining mechanical responses but also for controlling recrystallization processes[2,3]. For dealing with these, however, the widely-recognized dislocation density tensor, or equivalently, the concept of geometrically-necessary dislocations [4], seem to be incompetent, for it vanishes when counting both-signed components of dislocations simultaneously. This shortcoming brings about serious problems in tackling dislocation ensembles in general. The incompatibility tensor, on the other hand, has been demonstrated to readily and effectively solve those issues when it is appropriately utilized in the context of Field Theory of Multiscale Plasticity (FTMP) [5]. The theory relates the quantity with the energy-momentum tensor fluctuation after extended to 4D space-time [5], providing us with a new perspective regarding the quantity as the underlying microscopic degrees of freedom that plays critical roles as the destinations of excessively stored strain energy during the course of elasto-plastic deformations.

The present study targets four representative GNBs, whose detailed structures have been identified by Winther et al. [1]. The GNBs are reproduced by using dislocation dynamics method and the processes as well as the obtained configurations are evaluated based on FTMP, in particular, via duality diagram representations. We further characterize the simulated dislocation configurations by the Shannon entropy [6], and attempt to extensively examine in detail the organic interrelationships that can be expected to exist between the two quantities.

Theoretical Background

According to the non-Riemannian geometry, all the imperfections in crystalline materials space are completely represented by combinations of the torsion and the curvature [5]. The former is defined as the closure failure of the circuit around the defected field, while the latter is characterized by the rotation of the material vector after its parallel displacement along the

Explosion Shock Waves and High Strain Rate Phenomena
Materials Research Forum LLC
Materials Research Proceedings **13** (2019) 85-90
https://doi.org/10.21741/9781644900338-15

circuit. From the definition, they correspond to the dislocation density tensor and the incompatibility tensor, respectively.

The incompatibility tensor η_{ij} can be obtained as the spatial curl against the dislocation density tensor α_{ij} as,

$$\eta_{ij} = -\epsilon_{i|kl|}(\partial_{|k|}\alpha_{jl}) \tag{1}$$

In this study, we evaluate the dislocation density tensorαbased on the following equation.

$$b_l = \int_s \alpha_{jl} \, dS_j \tag{2}$$

Here, b_l is the Burgers vector and d$S_j$is the surface through which dislocation lines penetrate. We once obtain α_{ij} by dividing simulation cells into a prescribed number of sub-cells, and calculate the spatial derivatives.

In FTMP, presuming that incompatible displacements are driven by the inhomogeneous forces acting on them (flow-evolutionary hypothesis) [5], we derive the relationship between the incompatibility tensor and the fluctuation part of energy-momentum tensor δT_{ij}as,

$$\eta_{ij} = \kappa\delta T_{ij}, \tag{3}$$

where κ represents the duality coefficient, which is a sort of the transport coefficient,δmeans the deviation from the spatio-temporal average, and the subscripts i and j denote four-dimensional spatio-temporal components, i.e., i, j = 1, 2, 3, and 4.Considering the temporal components for the both tensors, we have a relationship between the incompatibility and the energy fluctuation, as follows.

$$\eta_{KK} = \kappa(\delta U^e + \delta K), \tag{3'}$$

where $U^e$is the elastic strain energy and $K$is the kinetic energy. In this study, K is defined as

$$K = \frac{1}{2}\rho b^3 l v^2, \tag{4}$$

where $\rho$is the density, b is the magnitude of Burgers vector, $l$is the segment length and $v$is the speed of dislocation nodes. In the static problem, K can be neglected. By plotting the two quantities in Eq.(3'), we obtain the duality diagram, allowing us to visualize the energy flow into the incompatibility-based degrees of freedom, accompanied by configurational changes occurring in the targeted dislocation systems.

The Shannon entropy $S_S^{config}$is evaluated as follows [6,7],

$$S_S^{config} = -\sum f_i \log f_i \tag{5}$$

Here,$f_i$is defined as the length ratio of the dislocation segments for the sub-cell i, i.e., the segment length of the dislocations contained in the sub-cell i divided by the total length within the simulation cell. Note that the commonly-used definition based on the number of points is not suitable in the present context, because the "number" varies as a function of the curvature of the segment independent of the segment length during the simulations.

Explosion Shock Waves and High Strain Rate Phenomena
Materials Research Proceedings **13** (2019) 85-90

Materials Research Forum LLC
https://doi.org/10.21741/9781644900338-15

Simulation Conditions and Results

After producing four representative GNBs [8], i.e., GNB2, 3, 4 and7, which has been identified and classified by Winther, et al. [1], we examine the formation processes based on the duality diagram representation scheme in the following.

Figure 1 summarizes the simulation results, displaying the initial and the final configurations, each compared with the ideal counterpart by Winther, et al. [1]. The GNB2 ultimately yields a regular hexagonal network, as shown in Fig.1, while relatively complex configurations accompanied by crooked segments are resulted for the GNB4. Mutually similar final configurations consisting basically of straight segments, on the other hand, are obtained for the GNBs 3 and 7. Generally, we confirm good agreements between the ideal morphologies and the corresponding simulated final configurations. To be noted that slightly curved segments in the simulated GNB4 are due to the residual internal stress field, not completely screened out even in the terminal state because of its highest dislocation density among others.

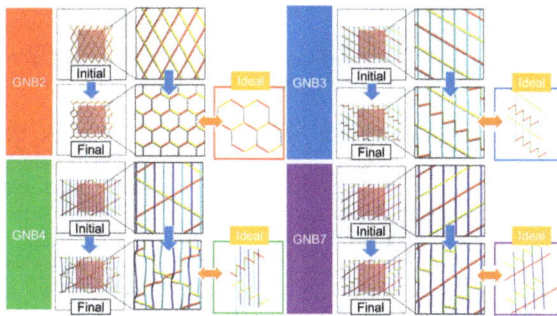

Fig.1 Simulated initial and final configurations for GNBs 2, 3, 4 and 7, compared with those predicted by Winther, et al [1].

Figure 2 displays duality diagrams for the above formation processes for the four GNBs, (a)without and (b)with considering the virtually-enhanced contributions of the kinetic energy fluctuation δK(eq.(3')), where the final states are indicated by solid symbols. The multiplying factor employed here is $\alpha = 1 \times 10^7$. As demonstrated in Fig.2(a), we find a unified trend governing all the GNBs: all the final states tend to align on a single master curve in the diagram. The virtual contributions of δK in Fig.3(b), on the other hand, indicate the rate of the configurational changes toward the final states, with the GNB2 exhibits the maximum, the GNB3 and 7 the intermediate, and the GNB4 yields the minimum, from which we can conclude the following. The GNB2 tends to reach its final configuration quite rapidly compared to others, as schematically illustrated in the insets in Fig.3(b) via rolling-down ball in steeper energy landscape.

The authors [8] proposed a new stability/instability criterion based on the above observations, as overdrawn on Fig.3(a), where the relationships between the disturbances$\Delta(\delta U^e)$ and the corresponding configurational changes measured by $\Delta\eta_{KK}$ for the GNB2 and GNB4 are indicated. The most stable GNB2 tends to cope with such disturbances by its relatively large configurational changes, whereas the GNB4 tries to withstand them rather firmly with minimum morphological variations. The above result about the virtual contributions ofδKsupports this postulate.

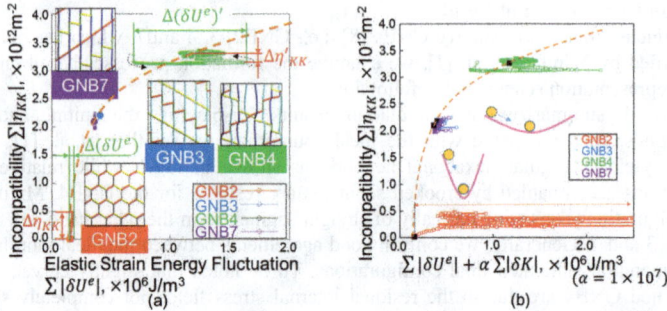

Fig.2 Duality diagrams for simulated GNBs (a) without and (b) with virtually enhanced contribution of kinetic energy term.

Relationships with Shannon Entropy

Since the Shannon entropy counts for the configurational information by nature, one can expect some correlations between what are to be represented by the quantity and the incompatibility discussed above.

Figure 3(a) shows the variation of the evaluated entropy with elapsed time for the four GNBs. The GNB2, with regular hexagon as the final configuration, exhibits the minimum entropy, together with the largest drop rate, while the GNB4, with the most complex configuration as well as the highest dislocation density among the four, yield the maximum value with the minimum mean decreasing rate, following relatively large temporal fluctuations. The GNBs 3 and 7, on the other hand, tend to converge to mutually similar values at the final configurations, simply because they basically have the same configuration. Figure 3(b) is the same plot as Fig.3(a) but for the incompatibility, from which we notice similar overall varying trend to that for the entropy. To further examine the similarity, we plot the entropy with the incompatibility in Fig.4(a). As expected, we confirm roughly a linear correlation between the two, where the trend seems to hold for both the initial and the final states, i.e.,

$$S_S^{config} = k^{config}\eta_{KK} \tag{6}$$

This strongly implies the incompatibility tensor contains also the features measured by the entropy, in particular, those about the configurational details.

Plotting the average rates of the two quantities, i.e., $\langle S_S^{config} \rangle_t$ with $\langle \eta_{KK} \rangle_t$ as shown in Fig.4(b), we also have a linear relationship between the two. Here, $\langle \ \rangle_t$ denotes the temporal average over the elapsed time until reaching each final configuration. This, together with the linear correlation between S_S^{config} and η_{KK} in Fig.4(a), implies,

$$\langle S_S^{config} \rangle_t = k^{config}\langle \eta_{KK} \rangle_t \tag{7}$$

with k^{config} representing the same coefficient as in Eq.(6).

Furthermore, we findsimilar trends also in the order of the mean decreasing rates $\langle S_S^{config} \rangle_t$ in Fig.3(b) and that of the magnitude of η_{KK} in Fig.3(b). In this regard, we additionally examine the relationship between the mean entropy rate and the incompatibility. Figure 5displays the plot of $\langle S_S^{config} \rangle_t$ versus η_{KK} on the semi-log basis. The mean entropy rate is demonstrated to be well correlated also with the incompatibility both at the initial and final states, which is equated as follows, i.e.,

$$\langle \dot{S}_S^{config} \rangle_t = k_{\ln}^{config} \ln \eta_{KK} \qquad (8)$$

The relarionship between Eqs.(7) and (8) deserves further investigations.

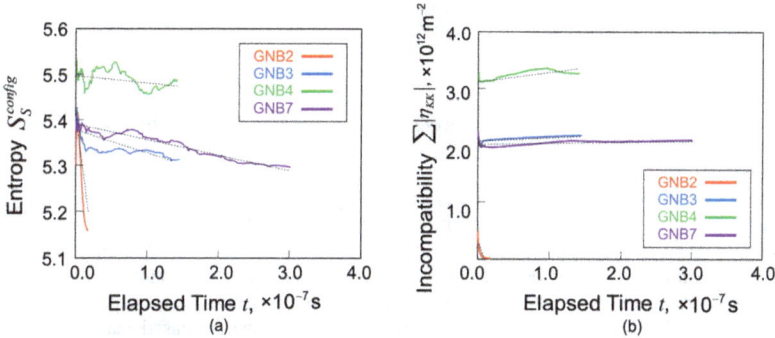

Fig.3 Variations of (a)entropy and (b) incompatibility for four GNBs.

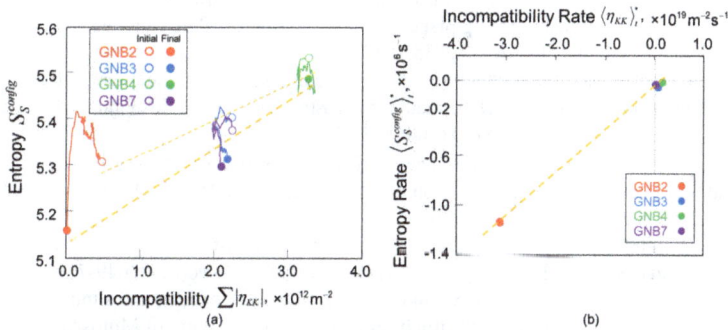

Fig.4 Relationships (a) between entropy and incompatibility and (b) between entropy rate and incompatibility rate.

Let us remind here the readers the flow-evolutionary hypothesis given by Eq.(3'), where the incompatibility is to be driven by the energy fluctuations. Since the entropy rate governs the rate of patterning in general, the linear relationships in Fig.4(b) and Fig.5 means that the dislocation patterning into GNBs is controlled by the energy fluctuations of the targeted systems.

In summary, those demonstrated above can provide a strong leverage that supports effectiveness of the duality diagram-based stability criterion proposed in [8], together with the critical roles played by the incompatibility tensor as the mediator for substantial understanding of the GNBs.

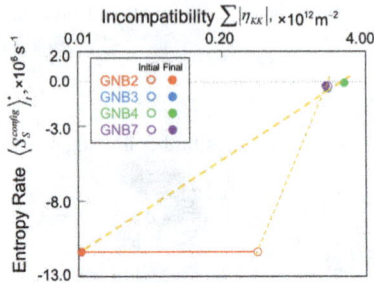

Fig.5 Relationships between entropy rate and log of incompatibility.

Summary

After briefly describing the FTMP-based evaluation against the four typical GNBs in the light of the duality diagram representation scheme, we further examine their configurational aspects based on the Shannon entropy. Extensive comparisons are made with the incompatibility tensor-based counterparts. The direct comparison of the two eloquently demonstrates that this simple conjecture seems to be effective, although its generality must be further examined. This finding also provides a leverage that strongly supports the duality diagram-based stability/instability criterion proposed in the previous study.

References

[1] G. Winther, C. Hong, and X. Huang, Low-Energy Dislocation Structure (LEDS) character of dislocation boundaries aligned with slip planes in rolled aluminium, Phil. Mag. 95 (2015) 1471-1489. https://doi.org/10.1080/14786435.2015.1033488

[2] N. Hansen, Deformation microstructures with a structural scale from the micrometre to the nanometre dimension, Evolution of Deformation Microstructures in 3D, Eds.Gundlach, C., et al. (Proc. 25th Riso Int. Symp Maters.Sci.) (2004) 13-32.

[3] O. V. Mishin, A. Gofrey, D. Juul Jensen and N. Hansen, Recovery and recrystalizaion in commercial purity aluminum cold roled to an ultrahigh strain, Acta Mater. 61 (2013) 5354-5364. https://doi.org/10.1016/j.actamat.2013.05.024

[4] Fleck, N.A., and Hutchinson, J.W., 2001, A reformulation of strain gradient plasticity, J. Mech. Phys. Solids 49, 2245-2271. https://doi.org/10.1016/s0022-5096(01)00049-7

[5] T. Hasebe, M. Sugiyma, H. Adachi, S. Fukutani, and M. Iida, Modeling and simulations of experimentally-observed dislocation substructures based on Field Theory of Multiscale Plasticity (FTMP) combined with TEM and EBSD-Wilkinson method for FCC and BCC poly/single crystals, Mater. Trans. 55 (2014)779-787. https://doi.org/10.2320/matertrans.m2013226

[6] Shannon, C.E., A mathematical theory of communication, Bell Syst. Tech. J. 27 (1948) 379-423.

[7] A.M. Hussein and J. A. El-Awady, Quntifying dislocation microstructure evolution and cyclic hardeining in fatigued face-centered cubic single crystals, J. Mech. Phys. Solids 91 (2016) 126-144. https://doi.org/10.1016/j.jmps.2016.03.012

[8] S. Ihara and T. Hasebe, FTMP-based simulations and evaluations of Geometrically-Necessary Boundaries (GNBs) of dislocation, *Model. Simul. Mater. Sci.* (Submitted) (2019).

Explosion Shock Waves and High Strain Rate Phenomena
Materials Research Proceedings 13 (2019) 91-96

Materials Research Forum LLC
https://doi.org/10.21741/9781644900338-16

Impact Joining of Metallic Sheets and Evaluation of its Performance

Minoru Yamashita[1, a *], Toshiki Shibuya[2] and Makoto Nikawa[1, b]

[1]Mechanical engineering, Gifu University, 1-1 Yanagido, Gifu, 501-1193, Japan

[2]Graduate student, ditto

[a]minoruy@gifu-u.ac.jp, [b]mniikawa@gifu-u.ac.jp

Keywords: Impact Joining, Mild Steel, Titanium, High-Speed Shear

Abstract. Similar or dissimilar metallic sheets were joined at their edges by the original impact joining method developed by one of the authors. Surface layers of both sheet edges activated by high-speed shear are immediately contacted with sliding motion in the joining process. The whole processing time is within a few milliseconds. The materials tested were mild steel and titanium sheets. Drop-weight impact testing machine was used. Joining performance of the fabricated sheets was evaluated by tensile test, etc. The joining was not available all over the thickness between sheets, in which sharp notch was observed near both sheet surfaces. The central portion was successfully joined without cavity. The joined specimen of mild steel and titanium was sliced to remove surfaces with such notch. Fracture occurs at the part of mild steel whose strength is lower, then the joining boundary was not damaged.

Introduction

It is well known that time and temperature effects have important role in solid state joining by atomic diffusion at elevated temperature. On the other hand, under cold condition, if the surface expansion is relatively large, two metal parts can join at the newly created surface, in which the brittle oxidized surface layer fractures. Joining strength in solid state welding was found to be approximately equal to the normal applied stress during the process in the absence of oxide films for the case of aluminum welded together in 1970 [1]. The film theory of such kind of welding or bonding was established, in which roll bonding was applied in 1983 [2]. Recently the film theory was used to derive a model that quantifies the relevance of these parameters to the weld strength [3]. Cold bonding may have a potential for recycling scrap aluminum [4].

The diffusion bonding is usually achieved by very high compressive stress with large plastic deformation. The shape drastically changes from the initial one and the joining strength also depends on the initial surface condition. Surface treatment is necessary for removal of the dirty surface layer. Experimental results in diffusion bonding were summarized for various metals including superplastic alloys [5]. Joining of different metals were tested [6] and experiments were carried out using super plastic materials [7, 8]. Hot isostatic pressing was also effective for the diffusion bonding of the nickel powder onto alumina tubing [9]. Divergent extrusion was used for bonding of aluminum by means of two opposing punches and finite element simulations was conducted [10]. However, the method requires very special conditions in temperature, atmosphere, surface treatment, etc. and they are very time consuming.

One of the authors proposed a novel joining method for sheet metal [11]. The edge of the sheet is joined to another edge, where the sheet thickness is unchanged, because the plates are not plastically compressed. In the present study, the materials are mild steel and pure titanium sheets. Main objectives are to observe the motion of the tools and the materials in the device, and to check the

Explosion Shock Waves and High Strain Rate Phenomena
Materials Research Proceedings **13** (2019) 91-96

Materials Research Forum LLC
https://doi.org/10.21741/9781644900338-16

deformation performance of the sheet composed of different materials by tensile and bending tests. The boundary was also inspected by elemental analysis.

Experimental device and materials tested

The impact joining setup is shown in Fig.1. The device is driven by an impact of drop-weight. The mass was 22 kg and the impact velocity was 10 m/s. The left half of the lower sheet is supported by the counter punch, whose reactive force is given by compressing the circular pipe (A6061, 12 mm diameter and 1 mm thick wall). The top edge of punch A is impacted then the simultaneous shearing commences. The upper sheared face slides to fit the lower sheared face. The motion stops at the prescribed position. The device is mounted on the low-elastic rubber that is pre-compressed to avoid the damaging excessive force.

Test materials are mild steel sheet SPC of 1.0 or 3.2 mm thickness, and pure titanium sheet TP340 of 1.0 or 3.0 mm thickness. Their tensile strengths were 303, 317 and 427, 401 MPa, respectively. Overlap length in sliding stage was varied.

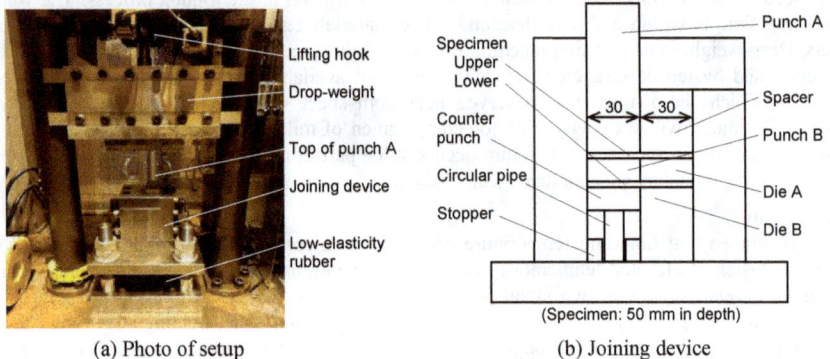

(a) Photo of setup   (b) Joining device

Fig.1 Experimental setup

Motion of tools and sheets in joining device

Motion of the tools and the sheets were observed with a high-speed video camera, in which the joining of TP340 (Upper specimen) and SPC (Lower one) was carried out. Progressive pictures are exhibited in Fig.2. The shear deformation and fracture of SPC are captured at time t: 87.7 and 439 μs, respectively. The left half of SPC moves downwards, then the TP340 also moves downwards after shear fracture. The TP340 appears at 2456 μs thereafter the edges of both materials slide each other with the prescribed overlap length at 3018 μs.

The sliding stage terminates at 3333 μs. The device sags due to the redundant energy of the drop-weight, where the low-elasticity rubber is compressed. It recovers at 6368 μs. Repulsion and contact between the tool and the drop-weight may occur in addition to the deceleration of the drop-weight, this causes the differences in moving distance of the tools calculated the impact velocity of 10 m/s.

Experimental result and discussions

Examples of joined specimens are exhibited in Fig.3 for the joining of different materials. No warping of the joined sheet with 1.0 mm thickness is observed, though the joining is not achieved all over the thickness. Sharp gap is observed at both surfaces of 1.0 and 3.0 mm thick sheets. The

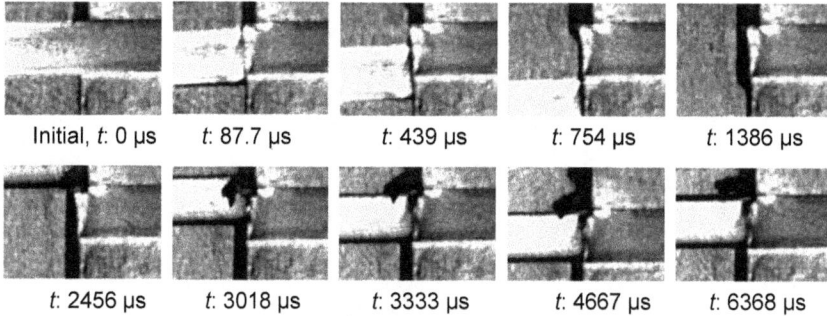

| Initial, t: 0 µs | t: 87.7 µs | t: 439 µs | t: 754 µs | t: 1386 µs |

| t: 2456 µs | t: 3018 µs | t: 3333 µs | t: 4667 µs | t: 6368 µs |

*Fig.2 Progressive photos of joining process of TP340 (Left) and SPC (Right)
observed by high-speed video (t: Time)*

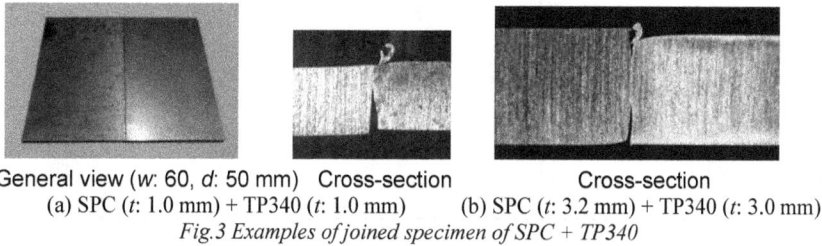

General view (w: 60, d: 50 mm)   Cross-section   Cross-section
(a) SPC (t: 1.0 mm) + TP340 (t: 1.0 mm)   (b) SPC (t: 3.2 mm) + TP340 (t: 3.0 mm)
Fig.3 Examples of joined specimen of SPC + TP340

joining is completed only for the central region. Protrusion of the SPC is seen at the upper gap. This is due to that TP340 scratches the softened surface material of SPC by high-speed shear. The sheared profile of thinner titanium sheet is not flat compared with the thicker one. This may be the reason of that the 1.0 mm thick titanium does not join.

Joining performance was evaluated as joining efficiency that is the relative tensile strength to the ultimate strength of the material. The specimen width is 10 mm. The performance is summarized in Fig.4 for the most appropriate overlap length L. It was set to 0.1, 0.2 or 0.3 mm. The performance is better for the thicker sheet in joining SPC together. The performance in joining TP340 together is also improved by increasing the sheet thickness or the sliding distance after shear process. Better performance in SPC is due to that the sheared face is flatter than that in TP340. However, the scattering in strength is remarkable. The reason is not clear at present, however, it may be attributable to the tool vibration, because the device is driven by a drop-weight.

The joining portion was specified by observing the boundary with a magnifying glass. Thickness of the specimen is decreased to approximately 0.5 mm by the removal of the both surface layers with sharp notch. The performance only for the apparent joined boundary is shown in Fig.4 (f) for 23 specimens. Over 90 % efficiency is obtained for 43 % of specimens. 100 % efficiency is achieved for 26 % of them, although the cases with very low efficiency also exist.

Figure 5 demonstrates the deformation patterns in tensile test. Initial width of the specimen was 3 ~ 5 mm. The maximum tensile stress can be determined appropriately, because the length is adequately long relative to the width. The specimen with 100 % efficiency exhibits diffuse necking at SPC part. It also fractures at the boundary in the latter case, where the width of SPC part shrinks.

Materials Research Forum LLC
https://doi.org/10.21741/9781644900338-16

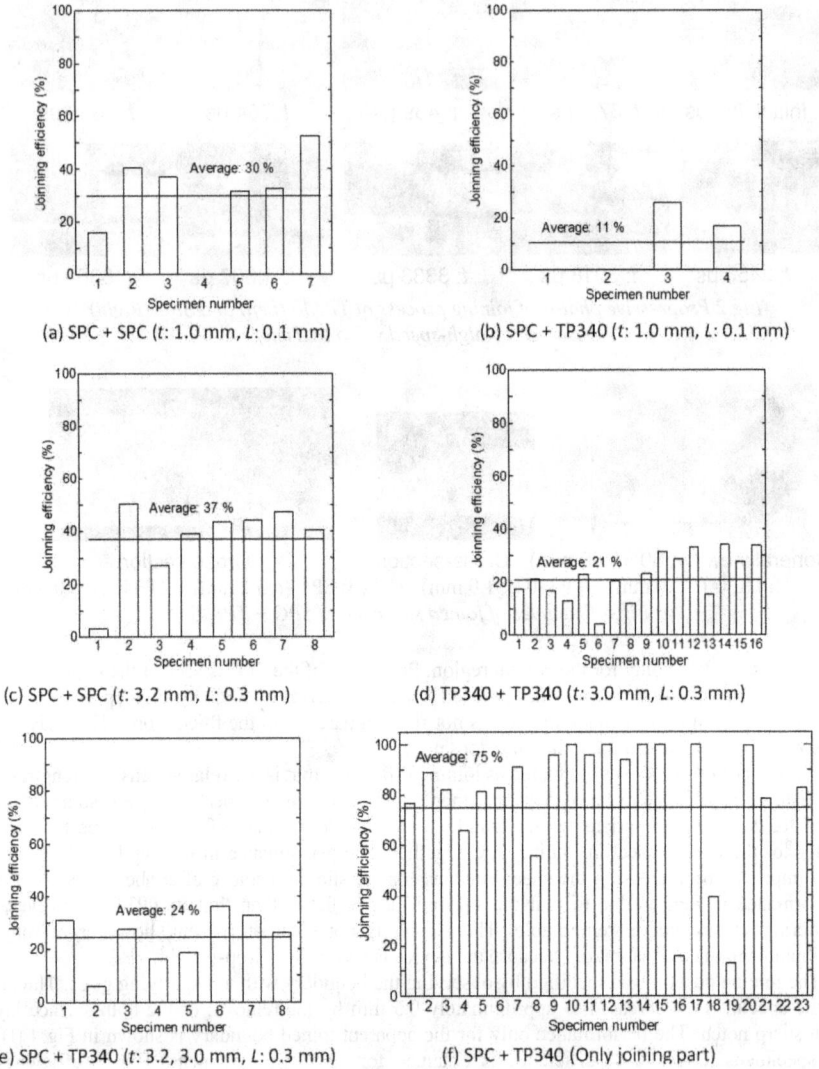

(a) SPC + SPC (t: 1.0 mm, L: 0.1 mm)

(b) SPC + TP340 (t: 1.0 mm, L: 0.1 mm)

(c) SPC + SPC (t: 3.2 mm, L: 0.3 mm)

(d) TP340 + TP340 (t: 3.0 mm, L: 0.3 mm)

(e) SPC + TP340 (t: 3.2, 3.0 mm, L: 0.3 mm)

(f) SPC + TP340 (Only joining part)

Fig.4 Summary of joining efficiency

The boundary of SPC and TP340 was analyzed with energy dispersive X-ray spectrometry as shown in Fig.6. Thin layer of a certain metallic compound was found to be generated. It may be Fe_2Ti or $FeTi$, which was found when titanium coating to the steel 35 by electro-spark deposition was carried out in argon [12].

Explosion Shock Waves and High Strain Rate Phenomena   Materials Research Forum LLC
Materials Research Proceedings **13** (2019) 91-96   https://doi.org/10.21741/9781644900338-16

(a) Initial

(b) Fracture at SPC

(c) Fracture at boundary
→ ← 5 mm

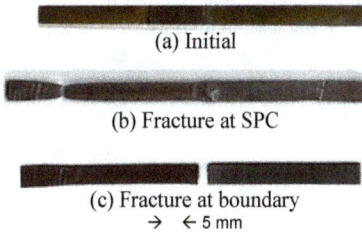

*Fig.5 Deformed shapes of tensile test specimen
(SPC + TP340, L: 0.3 mm)*

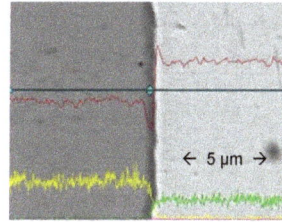

← 5 µm →

*Fig.6 Boundary of joined specimen
(Left: TP340, Right: SPC)*

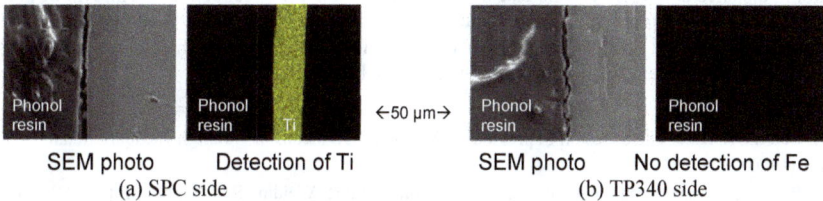

Phonol resin   Phonol resin   Ti   ←50 µm→   Phonol resin   Phonol resin

SEM photo   Detection of Ti   SEM photo   No detection of Fe
(a) SPC side   (b) TP340 side

Fig.7 Elemental analysis of notch portion (Specimen is enbeded in phenol resin.)

The surfaces in the notch are similarly analyzed as shown in Fig.7. The sheared face of the mild steel is covered with about 20 µm thick layer of titanium. This phenomenon reveals that the materials were once joined and separated during the sliding contact stage. However, the steel layer is not found at the surface of titanium. The volumetric heat capacity and the heat conductivity are lower in titanium. On the other hand, the yield stress is higher in the material. It implies that the surface layer of titanium is more softened than mild steel regardless that the melting temperature of titanium is about 200 K higher than steel.

Three-point bending test was conducted as shown in Fig.8. Width of the specimen is 25 mm. The span is 50 mm and the diameter of the anvils is 15 mm. Upper anvil is adhesively bonded to the sheet to avoid the relative slippage. The specimen SPC + TP340 of about 0.7 mm thickness is tested after removal of the surface layers with apparent notch. Figure 9 shows the 90° bent specimen, where the plastic deformation is very limited in the vicinity of the joining boundary. The joining was not achieved for the whole thickness. However, it is not separated. This suggests that the material holds ductility. For the narrower specimen shown in Fig.10, plastic deformation takes place

Fig.8 Three-point bending test   *Fig.9 Bent specimen and enlarged view of the boundary*

only in the SPC part that is weaker. The boundary is not damaged. The position of the upper anvil deviated from the central position at initial. The adhesive was not used in this case.

SPC   TP340

Joining boundary

Fig.10 Bent specimen (t: 0.8 mm, w: 3.8 mm)

Conclusions

Impact joining experiment was carried out for mild steel and titanium sheets. Observation of the motion of tools and sheets in the joining device by high-speed camera uncovers that the process terminates at about 3 ms. The joining performance is better for the thicker sheet attributed to that the sliding distance is longer. Tensile test was performed using the specimen after removal of both surface layers with sharp notch. The efficiency over 90 % is obtained for 43 % of test specimens, and that with 100 % was obtained for more than quarter, in which fracture with diffuse necking was observed at SPC part, there was no damage at the joining boundary.

References

[1] H. Conrad, L. Rice, The cohesion of previously fractured FCC metals in ultrahigh vacuum, Metall. Trans. 1 (1970) 3019-3029.

[2] N. Bay, Mechanisms producing metallic bonds in cold welding, Welding Res. Supplement (1983) 137-142.

[3] D.R. Cooper, J.M. Allwood, The influence of deformation conditions in solid-state aluminium welding processes on the resulting weld strength, J. Mater. Proc. Technol. 214 (2014) 2576-2592. https://doi.org/10.1016/j.jmatprotec.2014.04.018

[4] J.M. Allwood, Y. Huang, C.Y. Barlow, Recycling scrap aluminium by cold-bonding, Proc. 8th Int. Conf. Technol. Plasticity (2005) 311-312.

[5] H.Y. Wu, S. Lee, J.Y. Wang, Solid-state bonding of iron-based alloys, steel-brass, and aluminum alloys, J. Mater. Proc. Technol. 75 (1998) 173-179. https://doi.org/10.1016/s0924-0136(97)00323-3

[6] W. Elthalabawy, T.I. Khan, Diffusion bonding of austenitic stainless steel 316L to a magnesium alloy, Key Eng. Mater. 442 (2010) 26-33. https://doi.org/10.4028/www.scientific.net/kem.442.26

[7] H.S. Lee, J.H. Yoon, C.H. Park, Y.G. Ko, D.H. Shin, C.S. Lee, A study on diffusion bonding of superplastic Ti-6Al-4V ELI grade, J. Mater. Proc. Technol. 187-188 (2007) 526-529. https://doi.org/10.1016/j.jmatprotec.2006.11.215

[8] N. Ridley, Z.C. Wang, G.W. Lorimer, Diffusion bonding of dissimilar superplastic titanium alloys, Mater. Sci. Forum 243-245 (1997) 669-674. https://doi.org/10.4028/www.scientific.net/msf.243-245.669

[9] N.L. Loh, Y.L. Wu, K.A. Khor, Shear bond strength of nickel/alumina interfaces diffusion bonded by HIP, J. Mater. Proc. Technol. 37 (1993) 711-721. https://doi.org/10.1016/0924-0136(93)90130-x

[10] A. Lilleby, O. Grong, H. Hemmer, Experimental and finite element simulations of cold pressure welding of aluminium by divergent extrusion, Mater. Sci. Eng. A 527 (2009) 179-186. https://doi.org/10.1016/j.msea.2009.07.051

[11] M. Yamashita, T. Tezuka, T. Hattori, Impact joining of similar and dissimilar metal plates at their edges, Applied Mech. Mater. 566 (2014) 379-374.

[12] S. A. Pyachin, A.A. Burkov, Formation of intermetallic coatings by electrospark deposition of titanium and aluminum on a steel substrate, Surf. Eng. Appl. Electrochemistry 51 (2015) 118-124. https://doi.org/10.3103/s1068375515020131

Explosion Shock Waves and High Strain Rate Phenomena
Materials Research Proceedings 13 (2019) 97-102

Materials Research Forum LLC
https://doi.org/10.21741/9781644900338-17

Tensile Strength Test of Rock at High Strain Rate Using Digital Image Correlation

Tei Saburi[1, a *], Yoshiaki Takahashi[2,b], Shiro Kubota[1,c] and Yuji Ogata[2,d]

[1]National Institute of Advanced Industrial Science and Technology, Onogawa 16-1, Tsukuba, Ibaraki 305-8569, JAPAN

[a] t.saburi@aist.go.jp, [b] takahashi13r@mine.kyushu-u.ac.jp, [c]kubota.46@aist.go.jp, [d]yuji-ogata@aist.go.jp

Keywords: Tensile Strength, High Strain Rate, Blast Loading, Rock Fracture, Hopkinson Effect, Digital Image Correlation

Abstract. Tensile strength test of rock at high strain rate was experimentally performed by utilizing the nature of the strength difference. A magnitude of the tensile strength of brittle materials such as rock is much smaller than that of compressive strength. A compressive wave was produced by dynamic loading of explosive charge and made incident on a one end of a rock specimen bar. The compressive wave traveled through the specimen bar and it reflected at the free surface of the opposite end as a tensile wave with reversal amplitude. The tensile wave will cause the spall failure of the specimen at a specific distance from the free surface where the superposition of tensile and compressive waves exceeds the tensile failure strength of the specimen, usually referred to as Hopkinson effect. The dynamic behavior was observed at the side face of the bar specimen using a high-speed video camera, and the captured images were used to analyze the surface displacement behavior using a digital image correlation (DIC) technique. Strain and strain rate distributions on the specimen bar during impact loading were evaluated. The relationship between strain rate and dynamic tensile strength was discussed.

Introduction

Dynamic tensile strength is an important factor affecting rock fracturing and fragmentation during blasting operation in quarries and mines. For the dynamic strength test, the Split Hopkinson Pressure Bar (SHPB) is widely applied because of the wide range of strain rate applicability. Regarding the application of the SHPB method to brittle materials, there are many studies [1,2] such as concrete and rock materials for compressive strength. The SHPB method can be applied not only by the indirect tension [3] but also by the direct tension [4] for tensile strength. However, when the sample is rock, the pressure bars sandwiching the sample should be jointed even in the tension state. In the case of rock mass test materials, it is specified or recommended that the sample core diameter is 50 mm or more in ASTM [5] and 54 mm or more in ISRM [6] in indirect tension (Brazilian) test to secure the diameter of the material to some content from the presence of crystals and wrinkles. It is necessary to secure the diameter on the side of the incident bar and the transmission bar, and there is a concern that the system as the SHPB test device will become extensive. Therefore, we apply the dynamic tensile strength test using the Hopkinson effect in this study.

Explosion Shock Waves and High Strain Rate Phenomena   Materials Research Forum LLC
Materials Research Proceedings **13** (2019) 97-102   https://doi.org/10.21741/9781644900338-17

Experiments

The outline of the test equipment is shown in Fig.1. An explosive is placed on one end face of a cylindrical rock sample with a diameter of 30 mm. An explosive is detonated by the EBW detonator and impact pressure is applied to the sample. Thereby, a compressive stress wave propagates in the sample, and when it reaches the free end on the opposite side of the sample, it is reflected as a tensile stress wave. The area in which the reflected tensile stress wave propagates is an area that mixes with the compression wave that subsequently arrives. The net tensile condition of compressive stress minus tensile stress causes fracture at a point where the sample is sufficient to cleave and the sample leads to spalling. By evaluating the time change of the tensile stress calculated from the position and time of cleavage that occurred at this time and the displacement velocity that can be measured at the free end of the sample, the dynamic tensile strength can be estimated based on the distance and time from the free surface to the fracture surface. The displacement velocity on the free surface is measured by a laser Doppler displacement meter, and the specification of the fracture position and time is analyzed from a high-speed camera image. As the measuring instrument, a laser vibrometer LV-1610 (He-Ne 663 nm) from Ono Sokki was used for vibration measurement at the sample end. The high-speed camera SHIMADZU HPV-X (mono, 400x240 pixel) was used for the displacement measurement on the side generated by the stress wave propagating in the sample by the shock wave. By using two high-speed cameras, three-dimensional analysis of displacement is possible, but at present, distortion analysis in one axial direction was performed with one camera. The shooting was performed at a shooting speed of 500,000 fps. The stress propagating in the sample is evaluated from the strain information by strain gauge sticking, including the SHPB method. Although strain gauges can track only the time history of local strain at the attachment point, digital image correlation that performs strain measurement in a wide field of view by optical observation with the recent development of digital imaging technology and numerical calculation technology [7]. The scope of application of the Digital Image Correlation (DIC) method has been expanded, and by using a high-speed camera, it is possible to obtain dynamic strain distribution in a high strain range from the full field of view. Although observation using a high-speed camera has been performed in tensile fracture research using the Hopkinson effect [8,9], it has been used for dynamic distortion analysis in the entire field of vision, which is used only for judging the image of a fracture surface There are few examples. The behavior of rock material under dynamic tensile condition is analyzed and the relationship with tensile failure is clarified by vibration velocity measurement on the free surface of the sample by laser vibrometer and DIC strain analysis by high speed camera.

As a sample for the evaluation test, we used Isahaya sandstone produced in Nagasaki Prefecture. Table 1 shows typical mechanical properties. Generally, as in the case of rock materials, the tensile strength is much lower than the compressive strength, and it can be seen that the material is weak in tension. This test method is a test that utilizes the difference in compressive strength and tensile strength characteristic of rock materials.

Table 1 Material properties of Isahaya sandstone

Properties	Isahaya sandstone
Young's modulus, E [GPa]	32
Poisson ratio, ν [-]	0.39
Density, ρ [g/cm^3]	2.44
Uniaxial compressive strength, f_c [MPa]	182
Splitting tensile strength, f_{spu} [MPa]	9.9

Explosion Shock Waves and High Strain Rate Phenomena
Materials Research Proceedings **13** (2019) 97-102

Materials Research Forum LLC
https://doi.org/10.21741/9781644900338-17

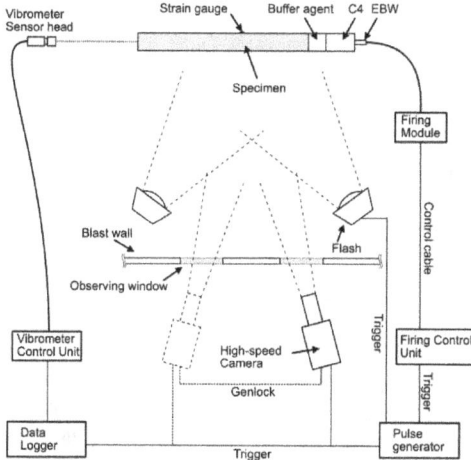

Fig.1 Schematic view of the experimental setup
(current system: one-camera for 2D-DIC and no buffer agent).

Fig. 2 Picture of sample speciman sprayed speckle patterns for DIC analysis.

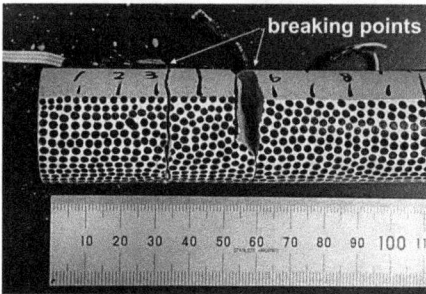

Fig. 3 Picture of sample after the shot.

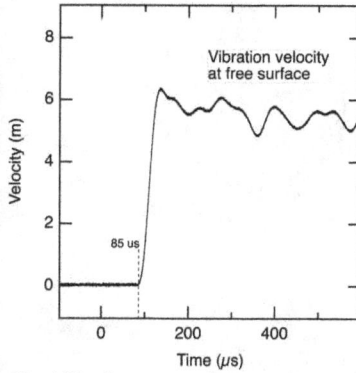

Fig.4 Displacement velocity at free surface.

Explosion Shock Waves and High Strain Rate Phenomena   Materials Research Forum LLC
Materials Research Proceedings **13** (2019) 97-102   https://doi.org/10.21741/9781644900338-17

Results and Discussion

The picture of the sample after the shot is shown in Fig. 3. Two spall fractures were identified at 33 mm and 55 mm from the free surface. The displacement velocity on the free surface is shown in Fig.4. The wave front reached the free surface after about 85 us, and the maximum displacement velocity thereafter reached about 6.35 m/s at about 135 us. The strain distribution in the sample area obtained by the DIC analysis of the captured images of the high-speed camera and the time history for every 10 us of the strain rate distribution are shown in Fig. 5 and Fig. 6, respectively. As a result of DIC analysis, we succeeded in visualizing the expansion and propagation of high strain area along with the propagation of compression wave by impact loading. In addition, it was observed that the compression wave reflected as a tensile wave at the free end, and this caused the strain distribution to be reversed and the fracture to be generated locally at a point where the tensile state became maximum. The strain and strain rate at break point reached up to 0.046 and 50 s^{-1}, respectively. Time histories of line profiles of strain along with the sample were extracted and were shown in Fig.7. It can be confirmed that the tension state starts to be predominant in the region between the two breaking points 95 us after the arrival of the reflected wave at the free surface. Time history of transmitting compression wave front was shown in Fig.8. The propagating velocity of the front was estimated from the slope of the plot and was estimated to be 3315 m/s, which is 40% lower than the elastic wave velocity calculated from the reference value shown in Table 1. Peak displacement velocity dU/dt at fracture point was evaluated as 1.83 m/s by DIC analysis which is relatively low compared with the half of the peak displacement velocity at free surface assuming free surface reflection as shown Fig. 4.

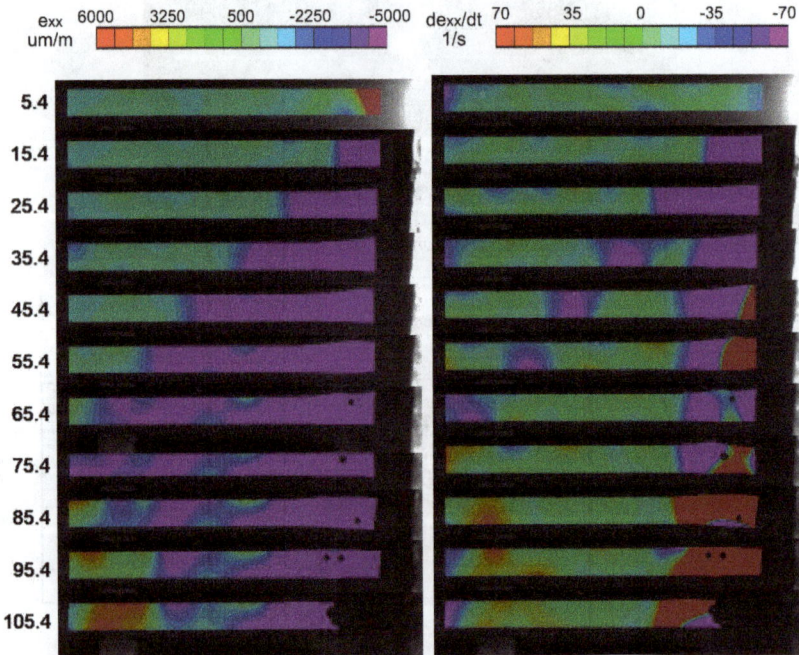

Fig.5 Sequences of strain distribution analyzed by DIC (e_{xx})

Fig.6 Sequences of strain rate distribution analyzed by DIC (de_{xx}/dt)

Explosion Shock Waves and High Strain Rate Phenomena   Materials Research Forum LLC
Materials Research Proceedings **13** (2019) 97-102   https://doi.org/10.21741/9781644900338-17

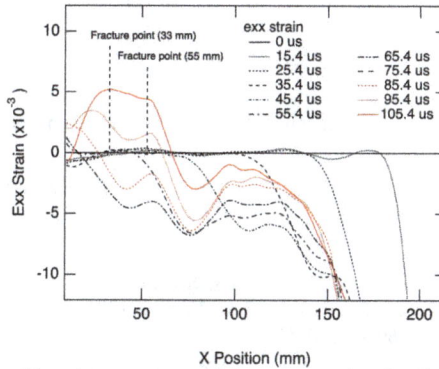

Fig.7 Line profiles of strain extracted from the results of DIC analysis (e_{xx}).

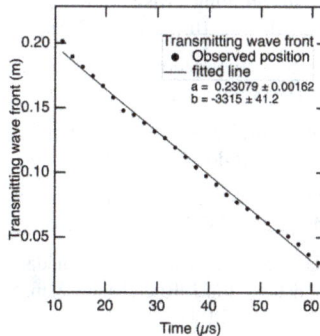

Fig.8 Time history of transmitting compression wave front.

Summary

In order to clarify the dynamic tensile failure behavior of rock material, dynamic uniaxial tensile experiment by the spalling test under impact loading of explosives based on the Hopkinson effect was proposed and was examined on a sandstone. The dynamic strain distribution and strain rate distribution on the sample were analyzed and visualized by Digital Image Correlation (DIC) method. The expansion and propagation of high strain area along with the propagation of compression wave were successively visualized. It was observed that the compression wave reflected as a tensile wave at the free end, and this caused the strain distribution to be reversed and the fracture to be generated locally at a point where the tensile state became maximum. The strain distribution and strain rate distribution were analyzed by applying DIC method. However, it is necessary to confirm the consistency with values obtained by other measurement methods and calculations. We will continue the experiment by changing the diameter, length and material of the specimen to clarify the relationship among material shapes, strain/strain rate distribution and tensile strength and to establish the quantitative estimation method for dynamic tensile strength of rock materials.

References

[1] C. A. Ross, J. W. Tedesco, S. T. Kuennen, Effects of strain rate on Concrete Strength, ACI Mat. J., 92 (1995) 37-47.

[2] Z. Zhou, X. Li, Z. Ye, K. Liu, Obtaining Constitutive Relationship for Rate-Dependent Rock in SHPB Tests, Rock Mech. Rock Eng., 43 (2010) 697-706. https://doi.org/10.1007/s00603-010-0096-3

[3] Q. B. Zhang, J.Zhao, A Review of Dynamic Experimental Techniques and Mechanical Behaviour of Rock Materials, Rock Mech. Rock Eng., 47 (2014) 1411-1478. https://doi.org/10.1007/s00603-013-0463-y

[4] S. Huang, R. Chen, K. W. Xia, Quantification of dynamic tensile parameters of rocks using a modified Kolsky tension bar apparatus, J. Rock Mech. Geotech. Eng., 2 (2010) 162-168. https://doi.org/10.3724/sp.j.1235.2010.00162

[5] ASTM, Standard Test Method for Splitting Tensile Strength of Intact Rock Core Specimens, ASTM D3967-08, (2008). https://doi.org/10.1520/d3967-95ar01

[6] ISRM, Suggested methods for determining tensile strength of rock materials, Int. J. Rock Mech. Min. Sci. Geomech. Abstr., 15 (1978) 99-103.

[7] M. A. Sutton, J. J. Orteu, H. W. Schreier, Image correlation for shape, motion and deformation measurements, Springer (2009). https://doi.org/10.1007/978-0-387-78747-3

[8] SH. Cho, Y. Ogata, K. Kaneko, Strain-rate dependence of the dynamic tensile strength of rock, Int. J. Rock Mech. Min. Sc., 40 (2003) 763-777.

[9] S. Kubota, Y. Ogata, Y. Wada, G.Simangunsong, H. Shimada, K. Matsui, Estimation of dynamic tensile strength of sandstone, Int. J. Rock Mech. Min. Sc., 45 (2008) 397-406. https://doi.org/10.1016/j.ijrmms.2007.07.003

Explosion Shock Waves and High Strain Rate Phenomena
Materials Research Proceedings **13** (2019) 103-108

Materials Research Forum LLC
https://doi.org/10.21741/9781644900338-18

Influence of Fiber Shape and Water-Binder Ratio on Blast Resistance of PVA Fiber Reinforced Mortar

Danny Triputra SETIAMANAH [1,a*], Makoto YAMAGUCHI[1,b], Priyo SUPROBO[2,c],
Shintaro MORISHIMA [1,d], Zicheng ZHANG[1,e],
Atsuhisa OGAWA [3,f] and Takashi KATAYAMA [3,g]

[1]2-39-1, Kurokami, Chuo-ku, Kumamoto-shi, Kumamoto 860-8555, JAPAN

[2]Raya ITS, Keputih, Sukolilo, Surabaya, East Java 60111, INDONESIA

[3]Ote Center Bldg., 1-1-3, Otemachi, Chiyoda-ku, Tokyo, 100-8115, Japan

[a]176d8938@st.kumamoto-u.ac.jp, [b]yamaguchi@arch.kumamoto-u.ac.jp, [c]priyo@ce.its.ac.id,
[d]184d9261@st.kumamoto-u.ac.jp, [e]172d8912@st.kumamoto-u.ac.jp,
[f]Atsuhisa.Ogawa@Kuraray.com, [g]Takashi.Katayama@Kuraray.com

Keywords: Fiber Reinforced Mortar, Contact Detonation, Local Failure, Fiber Shape, Water-Binder Ratio

Abstract. Reducing spall damage is a major problem when designing blast-resistant concrete structures. This study was conducted to evaluate the influence of various material factors on the blast resistance of FRCC slabs under contact detonation. The contact detonation tests were carried out on polyvinyl alcohol fiber reinforced mortar (PVAFRM) slabs with four different shapes of PVA fibers and four different water-binder ratios (W/B) of the mortar matrix. Fly ash (type II) was used as admixture and the fluidity of the PVAFRM in its fresh state was varied using a superplasticizer and thickener. As a result, it was obtained that longer fiber is more effective to suppress spall if the fiber diameter is constant, and if the aspect ratio of fiber (lf/df) is constant, finer fibers are more effective to reduce spall. Moreover, the spall-reducing performance is reduced when the W/B value is too high or too low, and it is considered that there is an appropriate value of W/B that depends on the fiber shape.

Introduction

When designing blast-resistant concrete structures, reducing spall damage is a major problem. Spalling indicates the failure of reinforced concrete (RC) slabs due to contact detonation which caused by the tensile stress waves reflected from the backside of the slab. To preserve human life under such circumstances, the launch of concrete fragments accompanies the spalling needs to be prevented. The authors have verified the good spall-reducing performance of fiber reinforced cementitious composite (FRCC) slabs under contact detonation[1]. However, a designing method for obtaining the required blast-resistant performance of the FRCC members has not been developed yet; one of the reasons for this is that it is difficult to obtain dynamic mechanical properties of FRCCs corresponding to this problem where the dominant strain rate is of the order of 10^3–10^4/s. Hence, it may be convenient to consider the spall-reducing performance of FRCC member as a material property of the FRCC. It can be obtained directly based on material factors such as fiber shape, water-binder ratio, and so on.

This study was conducted to evaluate the influence of various material factors on the blast resistance of FRCC slabs under contact detonation; therefore, contact detonation tests were carried out on polyvinyl alcohol fiber reinforced mortar (PVAFRM) slabs with four different shapes of PVA fibers and four different water-binder ratios of the mortar matrix.

Table 1 Materials used for PVAFRM.

Cement	Ordinary Portland cement; Density: 3.16 g/cm^3
Admixture	Fly ash (Type II); Density: 2.27 g/cm^3, Specific surface area: 3890 cm^2/g
Fine aggregate	Mountain sand; Surface-dried density: 2.56 g/cm^3, Water absorption: 2.29%, Maximum size: 2.5 mm, Fineness modulus: 2.58
Chemical admixture	Superplasticizer (Polycarboxylic-acid type); Thickener (Methylcellulose type)
Short fibers	PVA fiber (Type I); Density: 1.30 g/cm^3, Dimension: ϕ0.1 × 12 mm, Tensile strength: 1200 MPa, Tensile elastic modulus: 28 GPa PVA fiber (Type II); Density: 1.30 g/cm^3, Dimension: ϕ0.2 × 12 mm, Tensile strength: 975 MPa, Tensile elastic modulus: 27 GPa PVA fiber (Type III); Density: 1.30 g/cm^3, Dimension: ϕ0.2 × 18 mm, Tensile strength: 975 MPa, Tensile elastic modulus: 27 GPa PVA fiber (Type IV); Density: 1.30 g/cm^3, Dimension: ϕ0.2 × 24 mm, Tensile strength: 975 MPa, Tensile elastic modulus: 27 GPa

Table 2 Mixture proportions and static mechanical properties of PVAFRM.

Fiber type	V_f [%]	W/B [%]	FA/B [%]	S/B [%]	Unit weight [kg/m^3]					Sp/B [%]	Flow	γ [kN/m^3]	σ_B [MPa]	E [GPa]	ε_{co} [μ]	σ_f [MPa]	$\bar{\sigma}_b$ [MPa]
					C	FA	W	S	V								
I	2.0	50	20	100	649	162	406	812	0.5	0	205	18.7	35.2	14.4	4670	5.93	5.02
	2.0	40	20	100	707	177	353	883	0	0	155	20.3	57.5	20.4	5350	6.13	5.09
	2.0	33	20	100	753	188	311	942	0	0.6	177	19.4	57.4	20.7	4380	7.07	5.12
	2.0	25	20	100	815	204	255	1018	0	1.7	197	21.8	89.9	29.1	4480	9.70	7.24
II	2.0	50	20	100	649	162	406	812	0.5	0	257	19.4	39.2	15.8	4600	3.58	3.17
	2.0	40	20	100	707	177	353	883	0	0	212	20.6	61.0	20.1	5570	8.26	6.51
	2.0	33	20	100	753	188	311	942	0	0.5	217	20.9	73.5	23.6	5000	8.80	6.22
	2.0	25	20	100	815	204	255	1018	0	1.7	290	22.2	93.4	31.0	4670	7.69	6.00
III	2.0	50	20	100	649	162	406	812	0.5	0	238	19.3	37.3	15.5	4470	6.28	5.22
	2.0	40	20	100	707	177	353	883	0	0	212	20.6	57.6	19.7	4860	7.50	6.19
	2.0	33	20	100	753	188	311	942	0	0.5	215	19.5	57.8	20.8	4150	7.38	5.94
	2.0	25	20	100	815	204	255	1018	0	1.7	249	22.2	96.6	30.3	4620	9.90	8.16
IV	2.0	50	20	100	649	162	406	812	0.5	0	231	19.2	37.1	15.2	4670	5.78	4.71
	2.0	40	20	100	707	177	353	883	0	0	190	20.6	57.0	18.1	5140	9.15	7.81
	2.0	33	20	100	753	188	311	942	0	0.5	191	21.1	74.5	24.5	4730	10.1	7.05
	2.0	25	20	100	815	204	255	1018	0	1.7	260	22.2	92.8	28.3	5040	11.3	9.52

Notes; V_f: fiber volume fraction, C: cement, FA: fly ash, B (=C+FA): binder, W: water, S: sand, V: thickener, Sp: superplasticizer, γ: air-dried density, σ_B: compressive strength, E: Young's modulus, ε_{co}: strain at compressive strength, σ_f: flexural strength, and $\bar{\sigma}_b$: flexural toughness coefficient.

Materials

Table 1 lists the materials used for PVAFRM. The PVA fibers with the following shapes were employed: type I was ϕ0.1 × 12 mm (Aspect ratio (lf/df): 120); type II was ϕ0.2 × 12 mm (lf/df = 60); type III was ϕ0.2 × 18 mm (lf/df = 90) and; type IV was ϕ0.2 × 24 mm (lf/df = 120). To provide high fluidity to PVAFRM in its fresh state, fly ash (type II) was used as admixture.

Table 2 lists the mixture proportions of the PVAFRM. The water-binder ratio (W/B) varied over the values 25, 33, 40, and 50% with the fiber volume fraction, replacement ratio of cement by fly ash, and sand-binder ratio fixed at 2.0, 20, and 100%, respectively. The fluidity of the PVAFRM in its fresh state was varied using a superplasticizer and thickener. The static mechanical properties of various PVAFRMs are shown in Table 2.

Specimens Configuration

All the specimens were 500 mm long, 500 mm wide, and 80 mm thick, as shown in Fig. 1. It was investigated that the influence of the steel reinforcement on the local failure is negligible [2];

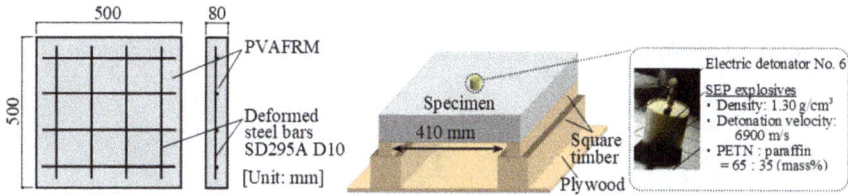

Fig. 1 Configuration ad bar-arrangement of speciment.   Fig. 2 Test set-up for contact detonation.

Table 3 Failure modes of PVAFRM specimens.

		Aspect ratio of PVA fiber			
		60 (d$_f$ = 0.2 mm, Type II)	90 (d$_f$ = 0.2 mm, Type III)	120 (d$_f$ = 0.2 mm, Type IV)	120 (d$_f$ = 0.1 mm, Type I)
W/B [%]	50	Spall	Spall	Spall	Crater
	40	Spall	Spall	Spall	Crater
	33	Spall	Crater	Crater	Crater
	25	Spall	Spall	Spall	Crater

however, lattice-like rebars with the horizontal and vertical intervals 120 mm in the center of slab thickness were employed to prevent the cracking of the slab.The PVAFRM was placed so that the casting surface became the blasted surface. These specimens were cured in water at 20 °C for 28 days, and then cured in air for 14 days until testing.

The specimen was supported by two wooden jigs with an inside span of 410 mm, as shown in Fig. 2. The SEP explosives were installed in the center of the upper surface of the specimen and blasted using an electric detonator No. 6. The amount of explosives is 70 g, which corresponds to the breach limit of a normal concrete slab with a thickness of 80 mm[2].

Fracture appearances
Tables 3 and 4 show the failure modes and the fracture appearances of the PVAFRM specimens, respectively. The following results can be obtained from these tables:
(a) Spall was prevented at W/B = 33% in types III and IV, whereas spall occurred at all W/B value in type II. Therefore, although it depends on W/B value, it is more effective to adapt longer fibers to suppress spall if the fiber diameter is constant; this may be because long fibers are difficult to be pull out.
(b) Spall was suppressed at all W/B value in type 1, but spall occurred at W/B = 50, 40, and 25% in type IV. Therefore, it is more effective to adapt finer fibers to suppress spall if the l_f/d_f value is constant. By analogy from the previous finding that rebar reinforcement hardly affects the spall failure[2], the reinforcing level becomes macroscopic by using thick-diameter fibers so that the effect of suppressing embrittlement of spall failure spot caused by dense cracks may be reduced.
(c) For types III and IV, spall was suppressed at W/B = 33%, but occurred at W/B = 50, 40, and 25%. Therefore, the spall-suppressing performance is reduced when the W/B value is too high or too low; it is considered that there is an appropriate value of W/B that depends on the fiber shape. In general, the lower the W/B value, the greater the bond strength at the fiber–matrix interface so that fiber pull-out is less likely to occur. On the contrary, as the W/B value increases, the pressure rise may be alleviated by the plastic compaction, which is a phenomenon in which the pressure rise is relieved by crushing the voids in the porous material. It is considered that the value of W/B is appropriate when the above two effects are balanced. As shown in Fig. 3, fine spall-fragments tend to increase in a low W/B mixture; the fiber-reinforcing effect may be reduced because the matrix was broken into small pieces by a strong inflation pressure.

Explosion Shock Waves and High Strain Rate Phenomena
Materials Research Proceedings **13** (2019) 103-108

Materials Research Forum LLC
https://doi.org/10.21741/9781644900338-18

Table 4 Fracture appearances of PVAFRM specimens subjected to contact detonation.

Notes; Each specimen was supported on both the left and right sides. The visible cracks that occurred on the blasted and rear sides were emphasized.

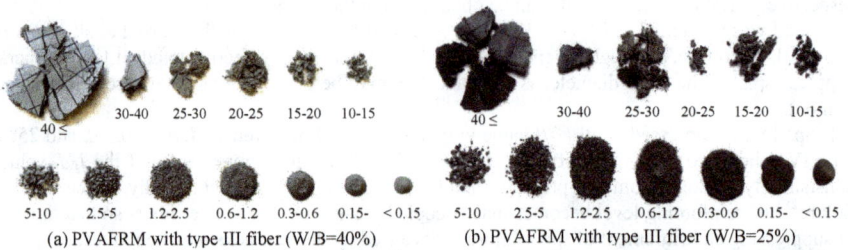

(a) PVAFRM with type III fiber (W/B=40%)　　(b) PVAFRM with type III fiber (W/B=25%)

Fig. 3 Examples of appearances of spall-fragments.　　　　[Unit: mm]

Local failure size

Figures 4 to 7 show the influences of l_f/d_f and W/B values on the local failure sizes. In these figures, the calculations by the equations for estimating the local failure sizes in a normal concrete slabs[2] are also shown.

As shown in Fig. 4, the crater diameter is almost constant regardless of the fiber shape, but the crater diameter tends to decrease with the decrease in W/B value. In the problem of contact detonation, it is expected that part of mortar at the crater portion will be extruded toward its outer peripheral part. According to this experimental results, because the bond at the fiber–matrix is strengthened as the W/B value decreases, the embrittlement of the outer periphery of the crater may be suppressed. However, the relationship between the crater diameter and the W/B value is not linear; the reducing effect tends to be moderate as the W/B value decreases.

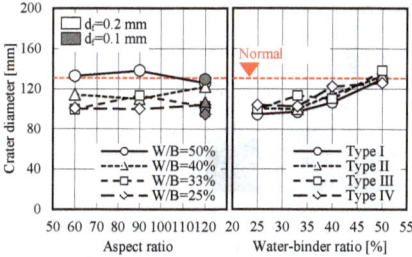

Fig. 4 Influence of l_f/d_f and W/B on crater diameter.

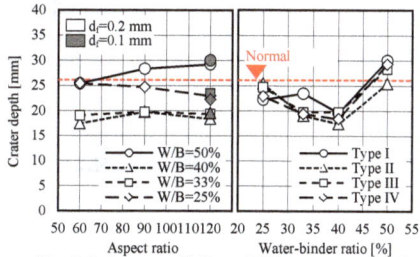

Fig. 5 Influence of l_f/d_f and W/B on crater depth.

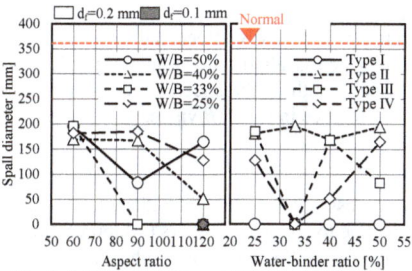

Fig. 6 Influence of l_f/d_f and W/B on spall diameter.

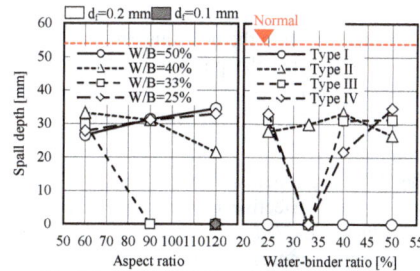

Fig. 7 Influence of l_f/d_f and W/B on spall depth.

On the other hand, it can be seen from Fig. 5 that the relationship between the crater depth and the W/B value is different from the above: the crater depth reaches a minimum value at $W/B = 40\%$. This may be because the crater bottom tends to be crushed due to the above-mentioned plastic compaction when $W/B = 50\%$; conversely, in the range where W/B value is 33% or less, because the plastic compaction effect is weakened with a decrease in W/B value, the crater bottom tends to be broken brittlely. Therefore, when the W/B value is higher or lower than its appropriate value, the spall occurs and the crater depth also increases, suggesting a risk that the breach easily occurs.

From Fig. 6, although the spall diameter is somewhat dispersed, it tends to be generally reduced as the fiber length increases, and no definite correlation with W/B value is observed. In addition, as shown in Fig. 7, the spall depth was within the range of 30 ± 4 mm in nine mixtures among ten mixtures in which spall occurred.

Influences of fiber

In general, it is known that fibers in FRCC tend to be two-dimensionally oriented in a plane parallel to the casting surface, except for the parts that are forcibly oriented by a formwork and so on. Therefore, it is predicted that the volume of fibers bridging the side part of the spall failure surface is high, and the volume of fibers bridging the top part is relatively less, as shown in Fig. 8 (a). As a result, the fibers bridging the side of the spall failure surface and the flexural cracking surface inside the spall-fragments mainly contribute to the prevention of spall (Fig. 8 (b)), and spall occurs when the bridging force is lost (Fig. 8 (c)).

(a) Fiber orientation inside the PVAFRM.

Fibers bridging at the side part of the spall failure surface and the flexural cracks inside the spall-fragments may contribute to the prevention of spall.

When the bridging force of fibers referred in (b) is lost, the spall may occur.

(b) Spall is suppressed.

(c) Spall occurs.

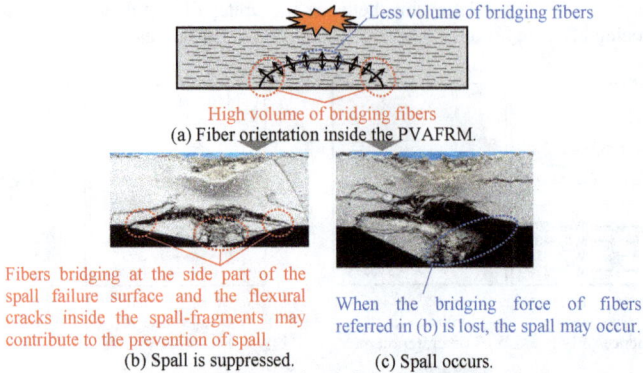

Fig. 8 Schematic image for influence of fiber

By analogy from the previous finding that concrete strength hardly affects the spall dimensions[2], it is expected that the position of the spall failure surface is almost invariable regardless of the matrix strength; therefore, it is considered that the spall depth of the specimen in which the spall occurred became almost constant irrespective of the difference in the W/B value.

Conclusion

The following conclusions were reached:
1) It is more effective to adapt longer fibers to suppress spall if the fiber diameter is constant. Further, if the l_f/d_f value is constant, it is more effective to adapt finer fibers to suppress spall.
2) The spall-suppressing performance is reduced when the W/B value is too high or too low, and it is considered that there is an appropriate value of W/B that depends on the fiber shape.
3) Although the crater diameter is reduced with the decrease in the W/B value, the optimum W/B value is attained when the crater depth reaches its minimum value.
4) In the specimens in which the spall occurred, the spall depth became almost constant even when the l_f/d_f and W/B values differed; this is possibility owing to the influence of the fiber orientation.

Acknowledgements

The contact detonation tests were carried out at the explosion laboratory of the Institute of Pulsed Power Science of Kumamoto University. The authors sincerely appreciate the help of Prof. K. Hokamoto, Assoc. Prof. N. Kawai, Assist. Prof. S. Tanaka, technical staff Y. Toda and A. Hamasaki, assistant technical staff T. Kusano, and students T. Ohmi, K. Gotoh, and K. Takasaki. The authors would also like to thank Kyuden Sangyo Co., Inc. and Shin-Etsu Chemical Co., Ltd. for their cooperation in conducting the experiments. This work was supported by JSPS KAKENHI (Grant-in-Aid for Scientific Research (C), Grant Number: 17K00647, Principal Investigator: M. Yamaguchi).

References

[1] M. Yamaguchi et al.: Blast Resistance of Polyethylene Fiber Reinforced Concrete to Contact Detonation, Journal of Advanced Concrete Technology, 9(1), pp.63-71, 2011. https://doi.org/10.3151/jact.9.63

[2] M. Morishita et al.: Effects of Concrete Strength and Reinforcing Clear Distance on the Damage of Reinforced Concrete Slabs Subjected to Contact Detonation, Concrete Research and Technology, 15(2), pp.89-98, 2004. https://doi.org/10.3151/crt1990.15.2_89

Explosion Shock Waves and High Strain Rate Phenomena
Materials Research Proceedings **13** (2019) 109-114

Materials Research Forum LLC
https://doi.org/10.21741/9781644900338-19

Effect of Pre-Notch on Deformation of Aluminium Square Plate under Free Blast Loading

Ali Arab[1,a*], P. Chen[1,b], Yansong Guo[1,c], Baoqiao Guo[1,d], Qiang Zhou[1,e]

[1] State Key Laboratory of Explosion Science and Technology, Beijing Institute of Technology, Beijing 100081, PR China

[a]arabali83@yahoo.com, [b]pwchen@bit.edu.cn, [c]gys009@qq.com, [d]baoqiao_guo@bit.edu.cn, [e]zqpcgm@gmail.com

Keywords: 3D DIC, Pre-Notch, Free Blast Loading

Abstract. In this paper the deformation and response of the Al square plate with pre-notch under free blasting loading were investigated. The blast loading process is a complex phenomenon, and accurate modeling, the prediction of deformation during this condition of loading is difficult, numerous investigations have carried out on the permanent deformation and the failure of various materials subjected to the blast; however, the transient deformation of plate subjected to the such intensely loaded has been difficult to measure due to the experimental difficulties. Recently, High speed imaging and 3D digital image correlation (3D DIC) have enabled the reliable none-contact measurement of the deformation and strain in various applications. In this research, 3D DIC method is used to study the effects of pre-notches on the dynamic deformation and rupture of thin Al square plates. The displacement and strain fields during the deformation were analyzed by this method. The Al square plate with the thickness of 3 mm and with different notch geometries were tested under the free blast loading by 20 gr of TNT. Two synchronized high-speed cameras were used to capture the response of the plate with an inter frame rate of 35000 frame/s. The result of test shows that the 3D DIC is a precise method to measure the full-field dynamic deformation of plate. The pre-notch possesses a direct effect on the maximum and final deformation of plate. Based on the result, the deformation occurs in the two stages; in the first one, the plate is deformed due to the shock impulse and all particles are forced to move out-of-plane and provided with initial velocities, and in the second one, pressure wave is vanished and deformation is occurred due to the imparted momentum.

Introduction

Explosion is a major threat for civil structure and vehicle in the combat zone. In recent years, many researches have carried out to improve the protection of the structure and vehicle against the blast. Measuring the deformation of plate subjected to the blast loading is one important task of solid mechanics. These studies can help for better understanding the dynamic response of the plate as well as identification and optimization of the influence design parameters under this harsh loading. However, most researches are only focused on the flat plate. Traditionally, the flat plate used in the structure as blast mitigation. The plate absorbs the explosion energy by either deformation or failure. Chen et al[1] studied the deformation of the circular aluminium plate under the confined blast loading through the 3D DIC technique. Markose and Rao[2] investigated the mechanical response of the V shape plate under the blast loading. They found that the geometry of plate has a direct effect on the mechanical response of plate. Kumar et al.[3] studied the effect of curvature on the response of plate subjected to the blast loading. Blast produces the shock wave that travelled through the target plate, and when it meets the other surface of the shock wave, gets reflected. The reflected shock wave on the flat surface is higher than incident shock considering the dissociation and ionization of real gases. This reveals that the resistance of plate against the blast is dependent

Explosion Shock Waves and High Strain Rate Phenomena
Materials Research Proceedings **13** (2019) 109-114

Materials Research Forum LLC
https://doi.org/10.21741/9781644900338-19

on its thickness[2]. However, effect of defect and crack on the response of the plate structure has not been investigated. In the combat situation defects and cracks could be occurred due to the fragmentation and combined blast. Rakvag et al.[4] studied the effect of various geometries of the preformed hole on the response of the steel plate subjected to the pressure pulse loading. However, in these tests, plates experience a large deformation and no localised deformation observed. Aune et al.[5] investigated the response of the steel plate with pre-formed hole subjected to the blast load. Liu et al.[6]carried out the blast experiments on the copper plate with pre-notch.

In this paper effect of pre-notch on the response of metal plate is investigated through 3D-DICmethod. The pre-notch with different size and geometry was prepared and tested under the free blast test.

Methodology
The layout of the experimental setup is shown in the fig.1 this facility is located in the explosion chamber located in the Xishian campus of the Beijing institute of technology. The plate is a flat square shape with size of 400*400 mm, targets assembled to the structure by the steel window frame. After assembly of the plate only 300*300mm of the plate is under the direct contact of blast shock wave. The pre-notch was made in the centre of plate by thickness 1mm and depth of 2mm. The details of experiments are shown in the table 1.

Before testing, surface of the plate was ground by sandpaper and washed by alcohol to remove all of the oil and dust. After cleaning the surface more than half of the surface of the targets was painted with white color and then speckled with black dots. This surface is used to perform high-speed digital image correlation (DIC) photography during the blast.

The shock wave is generated by the cylindrical shape TNT, the mass of explosive material and distance from the plate for each test is listed in the table 1. The explosion is ignited by the electrical detonator. The detonator ignite plus is also used to trigger the cameras.

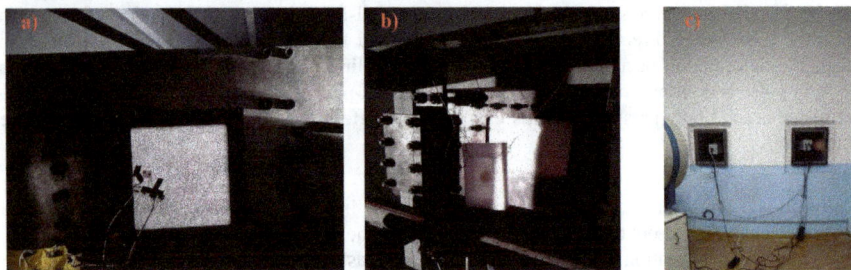

Fig 1. Experimental setup a) front view of the plate b) explosive material at back of the plate c) high speed cameras set up

Table 1. Dimension of tested samples

Sample name	material	Size of plate	Size of plate under the explosion loading	Explosion mass	Explosion distance	notch
Al-0	al	40*40cm	30*30	20 gr	22 cm	no
Al-30-0	al	40*40cm	30*30	20 gr	22 cm	Length =30mm-angle= 0 depth=2mm
Al-30-45	al	40*40cm	30*30	20 gr	22 cm	Length =30mm-angle= 45 depth=2mm
Al-40-0	al	40*40cm	30*30	20 gr	22 cm	Length =40mm-angle= 0 depth=2mm

3D-DIC

Digital image correlation is a non-contact optical method to study the motion, displacement and strain field. This method consists on capture consecutive images during applying the load to measure and evaluate the deformation and strain field. In this method the surface of the sample should paint by the random speckle pattern[7]. Details on principals and application of the 3D DIC can be obtained in the literature [8–11].

To analyse the 3D deformation of the plates subjected to the blast, two high speed cameras were used. These cameras were set outside of the explosion chamber to protect them from the shock wave of blast. They observe the same area of plate from the bullet proof windows. Cameras were linked and synchronized to each other by means of the synchronising cable that connected the cameras. The 35000 fps was set for both cameras with pixel of (320◊568). TTL trigger circuit was used to trigger the cameras and send pulse to start detonation. Three flash lamps were used to provide the enough illumination and avoiding the shading during the test. Before each test the position of the cameras was calibrated with calibration board (9*9 distance of 12). VIC-3D software was used to analyse the images and to calculate the deformation and strain of the plate. Based on the parameters of cameras and calibration coordination, by matching the reference images of the two cameras, the 3D surface contour of plate before blast can be reconstructed. And displacement and deformation of the plate can be calculated by the correlation, matching the deform images and reference image.

Result

As shown in the Fig. 2, a part of aluminium plate is selected as an area of interest (AOI) used to measure the displacement of the plate. Due to the limitation in the image resolution, the AOI is smaller than the whole plate size. Fig. 3(a) shows the evolution of aluminium plate without pre-notch (al-0) deformation along the vertical line in the centre of plate. After the detonation, the pressure wave travels through the air and arrives at the aluminium plate, creating a transient distributed load. Initially, this pressure is highly localized at the centre of plate. However, for the al-

Explosion Shock Waves and High Strain Rate Phenomena
Materials Research Forum LLC
Materials Research Proceedings **13** (2019) 109-114
https://doi.org/10.21741/9781644900338-19

0 plate, the maximum deformation in early stage does not occur in the centre of the plate, it could be due to the positioning of the charge, which may have had a small offset from the centre of plate. However, after0.25ms the maximum deformation is shifted to centre. Aune et al. [12] found that the geometry of the plate and boundary conditions had a greater influence than the positioning of the charge on the deformed shape. Fig. 3 (b) shows the out of plate displacement history of the several points on this line. Close to the border of the plate, particles are constrained due to the fixed boundary conditions and only have a limited initial displacement. During the blast loading, all particles of plate are forced to move out-of-plane (z- axis), when the load from the blast have vanished, the plate continues to deform. Spranghers et al.[13]found that the deformation history of plate consists of two parts. In the beginning, deformation of the plate is highly plastic and reaches a maximum value. After the plastic deformation, an elastic rebound occurs followed by elastic vibrations combined with damping[14]. Al-30-45 plate shows the maximum plastic deformation among the all plates. It is observed thatal-40-0 and al-0 plate experienced reversed snap buckling during the elastic rebound.

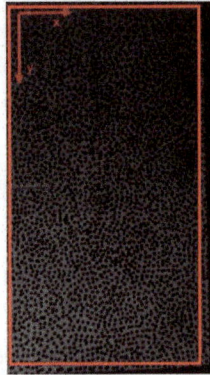

Fig.2 Area of interest (AOI)

Based on the DIC analysis, it can be seen that the maximum strain first occurs in the centre for the al-0 plate, and spreads rapidly outwards to the boundary. For the other plates the maximum strain was occurred at the pre-notch tip in early stage.

After the tests, all samples were collected for further analysis. There were no visible signs of tearing at the boundaries for all plates. For the al-30-45 and al-40-0 plates, it was observed that the cracks wereinitiated from the tips of the pre-notch and propagated. And also, for the al-40-0 plated, it was observed that rupture occurs at the location of pre-notch during the blast loading.

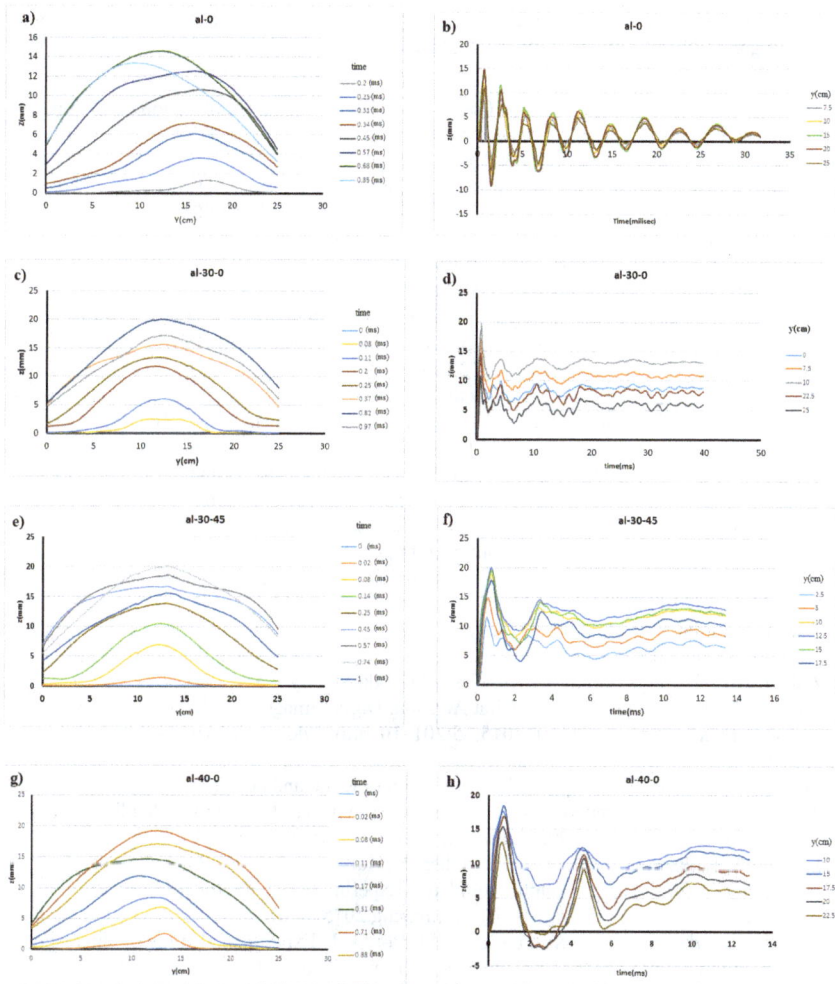

*Fig.3 evaluation of the displacement line profile along the vertical section line on the surfaces of
a)al-0, b)al-0,c)al-30-0, d)al-30-0, e)al-30-45, f)al-30-45, g)al-40-0 and h) al-40-0*

Summary

3D Digital Image Correlation (DIC) technique was applied to analyzing the response of pre-notched
plate subjected to the free air blast. The pre-notched plate experienced the larger plastic
deformations compare to the plate without notch. Failure was observed in al-30-45 and al-40-0

plate. The al-30-45 plate shows the greatest deformation among the all plates. This result shows the pre-notch has great influence on the response of plate to the blast loading.

References
[1]   Chen PW, Liu H, Ding YS, Guo BQ, Chen JJ, Liu HB. Dynamic Deformation of Clamped Circular Plates Subjected to Confined Blast Loading. Strain 2016;52:478–91. doi:10.1111/str.12190. https://doi.org/10.1111/str.12190

[2]   Markose A, Rao CL. Mechanical response of V shaped plates under blast loading. Thin Walled Struct 2017;115:12–20. https://doi.org/10.1016/j.tws.2017.02.002

[3]   Kumar P, Stargel DS, Shukla A. Effect of plate curvature on blast response of carbon composite panels. Compos Struct 2013;99:19–30. https://doi.org/10.1016/j.compstruct.2012.11.036

[4]   Rakvåg KG, Underwood NJ, Schleyer GK, Børvik T, Hopperstad OS. Transient pressure loading of clamped metallic plates with pre-formed holes. Int J Impact Eng 2013;53:44–55. https://doi.org/10.1016/j.compstruct.2012.11.036

[5]   Aune V, Valsamos G, Casadei F, Langseth M, B??rvik T. On the dynamic response of blast-loaded steel plates with and without pre-formed holes. Int J Impact Eng 2016;108:27–46. https://doi.org/10.1016/j.ijimpeng.2017.04.001

[6]   Liu H, Chen PW, Guo BQ, Zhang SL, Liu HB, An EF. Deformation and Failure of Pre-Notched Thin Metal Plates Subjected to Confined Blast Loading. Appl Mech Mater 2015;782:49–58. https://doi.org/10.4028/www.scientific.net/amm.782.49

[7]   Saburi T, Yoshida M, Kubota S. Dynamic strain distribution of FRP plate under blast loading. Int. Congr. High-Speed Imaging Photonics, vol. 10328, 2017, p. 103281N–6. https://doi.org/10.1117/12.2271148

[8]   Tiwari V, Sutton MA, McNeill SR. Assessment of high speed imaging systems for 2D and 3D deformation measurements: Methodology development and validation. Exp Mech 2007;47:561–79. https://doi.org/10.1007/s11340-006-9011-y

[9]   Hammer JT, Liutkus TJ, Seidt JD, Gilat A. Using Digital Image Correlation (DIC) in Dynamic Punch Tests. Exp Mech 2015;55:201–10. https://doi.org/10.1007/s11340-014-9924-9

[10]   Ren P, Zhou J, Tian A, Zhang W, Huang W. Experimental and numerical investigation of the dynamic behavior of clamped thin panel subjected to underwater impulsive loading. Lat Am J Solids Struct 2017;14:978–99. https://doi.org/10.1590/1679-78253521

[11]   Louar MA, Belkassem B, Ousji H, Spranghers K, Kakogiannis D, Pyl L, et al. Explosive driven shock tube loading of aluminium plates: Experimental study. Int J Impact Eng 2015;86:111–23. https://doi.org/10.1016/j.ijimpeng.2015.07.013

[12]   Aune V, Fagerholt E, Hauge KO, Langseth M, Børvik T. Experimental study on the response of thin aluminium and steel plates subjected to airblast loading. Int J Impact Eng 2016;90:106–21. https://doi.org/10.1016/j.ijimpeng.2015.11.017

[13]   Spranghers K, Vasilakos I, Lecompte D, Sol H, Vantomme J. Full-Field Deformation Measurements of Aluminum Plates Under Free Air Blast Loading. Exp Mech 2012;52:1371–84. https://doi.org/10.1007/s11340-012-9593-5

[14]   Curry RJ, Langdon GS. Transient response of steel plates subjected to close proximity explosive detonations in air. Int J Impact Eng 2017;102:102–16. https://doi.org/10.1016/j.ijimpeng.2016.12.004

Explosion Shock Waves and High Strain Rate Phenomena
Materials Research Proceedings **13** (2019) 115-120

Materials Research Forum LLC
https://doi.org/10.21741/9781644900338-20

Multiphysics Impact Analysis of Carbon Fiber Reinforced Polymer (CFRP) Shell

Cathrine Høgmo Strand[1,a], Zahra Andleeb[2,b], Hassan Abbas Khawaja[1,c*], Moji Moatamedi[3,d]

[1]UiT-The Arctic University of Norway

[2]Ghulam Ishaq Khan Institute of Engineering Sciences and Technology, Pakistan

[3]Oslo Metropolitan University, Norway

[a]cathrine_strand@hotmail.com, [b]gme1808@giki.edu.pk, [c]hassan.a.khawaja@uit.no, [d]mojtabam@oslomet.no

Keywords: Impact, CFRP, FEA, Cold Temperature

Abstract. With increasing popularity of Carbon Fiber Reinforced Polymer (CFRP) over time, the need for research in the field has increased dramatically. Many industries, i.e. aeronautical, automotive, and marine are opting to install carbon fiber in their structures to account for harsh environments like cold temperatures applications, but the research on the temperature exposure behavior of the materials are limited. This study aims to investigate the impact resistance of CFRP samples using the air gun tests. Two different shaped pellets (Diabolo and Storm pellets) were used in this work. The pellets speeds were calculated using a high-speed camera. The tests were performed in the room temperature (22°C) as well as in the cold room where the test pieces were exposed to about -28°C for seven days. The experimental studies were performed and compared against finite element simulations using ANSYS®. The studies also included layering of the CFRP samples to find the limiting thickness of pellets penetration. It was concluded that the thickness of 0.79mm and below of CFRP, cannot resist the impact of pellets. The visual inspection of failure revealed that the CFRP has gone through a brittle failure. However, temperature was found to have no significant impact on the results as similar behavior of CFRP was observed in both room conditions (22°C) and cold temperatures (-28°C).

Introduction

In the last decades, a growing interest has been dedicated in the use of composite materials for structural applications. CFRP composites are gaining a special attention to replace traditional materials in several fields although it is well known that these systems are highly susceptible to internal damage caused by transverse loads even under low-velocity ones [1,2]. In general, CFRP composites can be damaged on the surface and also beneath the surface by relatively light impacts causing invisible impact damage [3]. Therefore, this study has been carried out both to highlight effects of variables linked to geometrical parameters of composite sheets, impactor, and operative conditions. Therefore, this study has been carried out both to highlight effects of variables linked to geometrical parameters of composite sheets, impactor, and operative conditions. Operative conditions affect the material properties as reported in [4-6].

Experimental Setup

a. Test Samples

Test samples used in this study were from the DragonPlate®, manufactured by Allred and Associates Inc., Elbridge, New York [7]. The CFRP samples used were EconomyPlate™ Solid

Carbon Fiber Sheet ~ 1/32" x 12" x 12" (0.79375mm x 304.8mm x 304.8mm) [8]. EconomyPlate™ sheets comprised of orthotropic (non-quasi-isotropic) at 0°/90° orientation laminates [9] (Figure 1.) utilizing a twill weave [10] (Figure 2.), while maintaining a symmetrical and balanced laminate. EconomyPlate™ composed entirely of a tough and rigid carbon reinforced epoxy matrix, with textured finish on both sides. Samples were cut into smaller pieces for test purposes (Figure 3.).

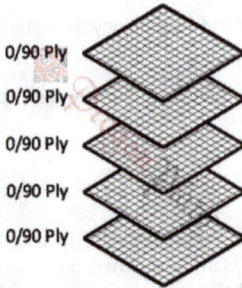

Figure 1-0°/90° orientation laminate

Figure 2-Carbon fiber twill weave

Figure 3-CFRP test samples

b. Impact Tests

To perform the impact tests, a shooting box was built, as shown in Figure 4a. The shooting box was designed such that it collects the pallets once they pass through the samples. The box consists of an opening-closing system with locking screws and wingnuts, so test pieces could be fastened for testing (Figure 4b), and removed and replaced with new test pieces effectively. Diabolo and storm pellets (Figures 5 and 6.) were shot on to the CFRP test samples. The material of both pellets was lead and they were of 4.5mm caliber, weighing about 0.5g each. The test was performed in room temperature, on tempered test pieces at about 22°C and in the cold room on test pieces exposed to about -28°C for 7 days.

Materials Research Forum LLC
https://doi.org/10.21741/9781644900338-20

(a) Shooting box

(b) Fastened test piece

Figure 4 – The opening-closing system of the shooting box

Figure 5-Diabolo pellets

Figure 6-Storm pellets

A speed tests were carried out using a high-speed camera (Figure 7.). The test showed the pellets speed of about 160m/s.

Figure 7-Speed test of Diabolo pellet (pellet speed ~ 160m/s)

Experimental Results

Impact tests revealed that diabolo and storm pellets at 160m/s pass through the single layer (~0.79 mm) of CFRP (Figure 8). Visual inspection showed that the CFRP test samples were ruptured (brittle failure) and the failure was in the close vicinity of the impact. Ruptured holes were more visible when Storm pellets were used, nonetheless, the failure areas were the same.

(a) Diabolo pellets   *(b) Storm pellets*

Figure 8-Visual inspection of the impact

Tests were repeated by tightly joining the layers of CFRP tests samples (0.79mm, 1.59mm, and 2.38mm). Pellets passed through 0.79mm and 1.59mm thick CFRP test samples, however, deflected for 2.38mm layer. Same results were observed when tests were conducted at room temperature conditions (22°C) and cold conditions (-28°C).

Simulations Setup

The simulations were performed in ANSYS® Explicit Dynamic [11]. Mesh sensitivity analysis was performed to ensure the accuracy of results. The model parameters are given in Table 1.

Table 1: Simulation model parameters (ANSYS® Explicit Dynamic)

Physics preference	Explicit
Relevance	70
Relevance Center	Fine
Span Angle Center	Fine
Nodes (optimized)	9193
Elements (optimized)	13786

Simulations Results

ANSYS® Explicit Dynamic simulations revealed similar behavior as seen in experiments. For example, 0.79mm CFRP went through rupture failure as shown in Figure 9. Tsai-Wu failure model was used in the simulations [12].

Explosion Shock Waves and High Strain Rate Phenomena
Materials Research Proceedings **13** (2019) 115-120

Materials Research Forum LLC
https://doi.org/10.21741/9781644900338-20

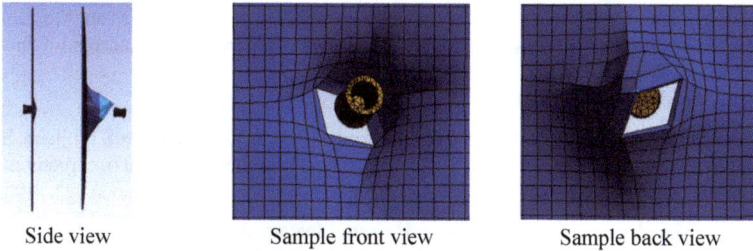

| Side view | Sample front view | Sample back view |

Figure 9- ANSYS® Explicit Dynamic simulations

Comparison of Experiments and Simulations

Table 2 summarizes the results from experiments and simulations. As shown,

	Experiments	Simulations
CFRP thickness = 0.79mm @ 25°C to -28°C	Failed	Failed
CFRP thickness = 1.59mm @ 25°C to -28°C	Failed	Failed
CFRP thickness = 1.63mm @ 25°C to -28°C	(not tested)	Safe
CFRP thickness = 2.38mm @ 25°C to -28°C	Safe	Safe

Conclusions and Limitations

Following conclusion can be drawn from the study:

1. It can be concluded that pellet and storm pellets at 160 m/s can damage/pass through the 1.59mm and below thickness of CFRP.
2. Good agreement was found between the experiments and simulations. It confirms that Multiphysics methodology such as Explicit Dynamic simulations may be used for the design of CFRP structures undergoing impact loading.
3. It was found that CFRP material properties did not change noticeably in cold temperatures.

Following limitations apply to the given study:

1. Commercially available CFRP samples (DragonPlate®) were used in this study.
2. Commercially available Multiphysics software ANSYS® was used for the simulations.
3. Samples were visually inspected and not for micro-fractures/micro-delamination.

Acknowledgement

Thanks to Prof. Young Kwon from Naval Postgraduate School, Monterey, California, USA for providing the test samples.

References

[1] Khawaja, Hassan Abbas; Moatamedi, Mojtaba. Multiphysics Investigation of Composite Shell Structures Subjected to Water Shock Wave Impact in Petroleum Industry. Materials Science Forum 2013. doi: https://doi.org/10.4028/www.scientific.net/MSF.767.60.

[2] Khawaja, Hassan Abbas; Messahel, Ramzi; Souli, Mhmed; Al-Bahkali, Essam; Moatamedi, Mojtaba. Fluid solid interaction simulation of CFRP shell structure. Mathematics in Engineering, Science and Aerospace (MESA) 2017, 8(3), p. 311-324. Link: http://nonlinearstudies.com/index.php/mesa/article/view/1532

[3] Khawaja, Hassan Abbas; Bertelsen, Tommy; Andreassen, Roar; Moatamedi, Mojtaba. Study of CRFP Shell Structures under Dynamic Loading in Shock Tube Setup. Journal of Structures 2014, doi: http://dx.doi.org/10.1155/2014/487809.

[4] Stange, Even; Andleeb, Zahra; Khawaja, Hassan; Moatamedi, Mojtaba. Multiphysics Study of Tensile Testing using Infrared thermography. The International Journal of Multiphysics 2019; 13(2), p. 191-202. doi: http://dx.doi.org/10.21152/1750-9548.13.2.191

[5] Myrli, Odd Einar; Khawaja, Hassan. Fluid-Structure Interaction (FSI) Modelling of Aquaculture Net Cage. The International Journal of Multiphysics 2019; 13(1). p. 97-111. doi: http://dx.doi.org/10.21152/1750-9548.13.1.97

[6] Ahmad, Tanveer; Khawaja, Hassan. Review of Low-Temperature Crack (LTC) Developments in Asphalt Pavements. The International Journal of Multiphysics 2018; 12(2). p. 169-187. doi: http://dx.doi.org/10.21152/1750-9548.12.2.169

[7] Allred and Associates Inc - Company. [cited 03.03.2019]; Available from: http://dragonplate.com/sections/company.asp.

[8] Allred and Associates Inc - Product. [cited 03.03.2019]; Available from: https://dragonplate.com/economyplate-solid-carbon-fiber-sheet-1_32-x-12-x-12.

[9] Allred and Associates Inc – Non-quasi-isotropic. [cited 03.03.2019]; Available from: https://dragonplate.com/quasi-isotropic-carbon-fiber-sheets.

[10] Allred and Associates Inc -Twill weave. [cited 03.03.2019]; Available from: https://dragonplate.com/what-is-carbon-fiber.

[11] ANSYS® Explicit Dynamic - [cited 03.03.2019] Available from: https://www.ansys.com/products/structures/ansys-explicit-dynamics-str.

[12] Tsai, Stephen; Wu, Edward. A general theory of strength for anisotropic materials. Journal of Composite Materials 1971. 5(1) p. 58-80. doi: https://doi.org/10.1177%2F002199837100500106

Explosion Shock Waves and High Strain Rate Phenomena
Materials Research Proceedings **13** (2019) 121-127

Materials Research Forum LLC
https://doi.org/10.21741/9781644900338-21

ANFIS Modeling for Prediction of Particle Size in Nozzle Assisted Solvent-Antisolvent Process for Making Ultrafine CL-20 Explosive

Dinesh K. Pal[1,a*], Shallu Gupta[1,b], Deepika Jindal[1,c], Anil Kumar[2,d], Arun Agarwal[1,e] and, Prem Lata[3,f]

[1]Terminal Ballistics Research Laboratory, DRDO, Sector-30 D, Chandigarh-160030, India

[2]University Institute of Engineering and Technology, Panjab University, Hoshiarpur, India

[3]Department of Applied Sciences, PEC University of Technology, Chandigarh-160 014, India

[a]dineshpalb@yahoo.com, [b]shallugupta74@yahoo.com, [c]frn.deepika@gmail.com, [d]eed.tbrl@gmail.com, [e]arun907@yahoo.com, [f]prem_lata36@yahoo.com

Keywords: Ultrafine CL-20, Artificial Neural Network, ANFIS, Neuro-Fuzzy

Abstract. Physical properties such as particle size, surface area and shape of explosive control the rapidity and reliability of initiation, and detonation and thus determine the performance of an explosive device such as slapper detonators. In this paper, Nozzle assisted solvent/antisolvent (NASAS) process for recrystallisation of CL-20 explosive is established. Many process parameters are involved which affect the particle size of the explosive. Therefore an accurate prediction of particle size is required to tailor the particle size. In the present work, an intelligent algorithm is applied to build a simplified relationship between recrystallization process parameters and particle size. This can be used to predict explosive particle size with a wide range of process parameters through an adaptive neuro-fuzzy inference system (ANFIS). The model is trained using experimental data obtained from design of experiment techniques utilizing a MATLAB software. Six process parameters such as Solution pressure, Antisolvent pressure, Antisolvent temperature, Stirrer speed, Solution concentration and Nozzle diameter are used as input variables of the model and the particle size is used as the output variable. The predicted results are in close agreement with experimental values and the accuracy of the model has been tested by comparing the simulated data with actual data from the explosive recrystallization experiments and found to be inacceptable range with maximum absolute percentage error of 11.52 %. The ultrafine CL-20 prepared by NASAS process is used in Slapper detonator application. The threshold initiation voltages for CL-20 based slapper detonator is found to be in the range of 0.9 kV with standard deviation of ±0.1 kV.

Introduction

The physical properties such as crystal particle size, shape, morphology, crystalline imperfections, purity and microstructure of the inter-crystalline voids of an existing explosive can be altered. There are wide variety of processes available for tailoring particle size and morphology of energetic materials such as solvent/non-solvent recrystallization[1],continuous crystallization of submicrometer energetic materials [2], spray flash evaporation [3]Yang et al. [4] obtained nano-TATB by using solvent/anti-solvent method with a particle size of 60 nm approximately through atomization of solution by a nozzle to small droplets and colliding rapidly with non-solvent flow. There is a need of mathematical model to predict particle characteristics as a function of process parameters to provide a basis for a computer based process control system. Shallu Gupta et al.[5,6], used micro nozzle assisted spraying process (MNASP) for recrystallization of Submicrometer Hexanitrostilbene (sm-HNS) Explosive. The process attributes were optimized using weighted average techniques of Analytical Network Process (ANP). The advantages of neural network based

Materials Research Forum LLC
https://doi.org/10.21741/9781644900338-21

techniques include extreme computation, powerful memory and rapid learning from experimental data. Furthermore, it can predict an output parameter with accuracy even if the input parameter interactions are not completely understood[7, 8]. *Artificial neural network (ANN)* and multilayer perceptron (MLP) is widely established inartificial intelligence (AI) research where a nonlinear mapping between input and output parameters is required for a function approximation[9, 10]. Pannier *et. al*, have explained the application and general features of Fuzzy logic (FL)modeling, fuzzy sets, membership functions, and fuzzy clustering[11]. theoretical details of the neuro-fuzzy modeling can be found in [12, 13]. Moreover, however, relevant features and context that refer to the adopted means of neuro-fuzzy modeling, i.e., ANFIS [14]

It is seen from literature that in spite of being powerful modeling tool, ANFIS has not been used in the study of explosive recrystallization process. A neuro-fuzzy technique called *adaptive network based fuzzy inference system* (ANFIS) combines fuzzy systems with neural networks, utilizing the learning characteristics of neural network and decision making capability of fuzzy systems. In this research work, application of ANFIS model is adopted for predicting the particle size of CL-20 explosive in solvent-antisolvent recrystallization process.

Experimental Work

The explosive material used in this research work was raw ε-CL-20 with a particle size in the range of 50 to 60 μm. In this research work, for making UF-CL20, a Nozzle Assisted Solvent-Antisolvent (NASAS) process has been designed, developed, fabricated and installed, as per schematic diagram shown in Fig. 2. The NASAS process was used to carry out 49 experiments for making UF-CL20 explosive. Based on design of experiments, six input parameters were considered which affect the output of the re-crystallized explosive i.e. particle size. The input parameters are - solution pressure, anti-solvent pressure, anti-solvent temperature, stirrer speed, solution concentration and nozzle diameter. The output parameter i.e. particle size was used as the response variable. The UF-CL20 obtained by NASAS process was characterized as explained in the following section.

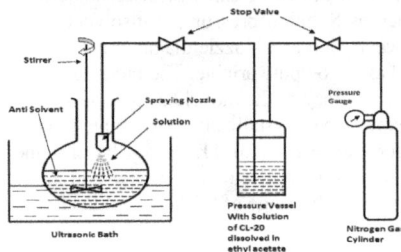

Figure 1. *Schematic of NASAS process*

Characterization

The distribution of particle size for some of the samples under similar condition is shown in Fig. 3 with mean particle size of UF-CL20 as 2.61 μm with standard deviation of 0.242 μm. *Total 42 Nos. of experiments were carried out to record the 42 data of input-output pairs of variables shown in Table 1 for ANFIS model.* Recrystallised ultrafine CL-20 was characterized using XRD analysisto ensure crystalline nature XRD pattern showed the peaks at similar difraction angle as those of CL-20 which exhibits a unique non-overlapping diffraction peak at 19.98 2θ, as shown in Fig.4. FTIR analysis was carried out to ensure there is no change in molecular structure after processing as shown in Fig.5. SEM photography showed the reduction of particle size and the morphology was also affected by the process parameters as shown in Fig. 6. The shape is a mix of polyhedral and nearly spherical geometry. The surface seems to be smooth with negligible defects/ cracks.

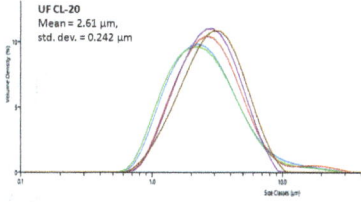

Figure 2. *Particle size distribution*

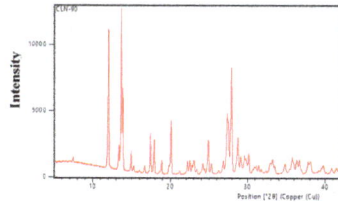

Figure 3. *XRD pattern of processed CL20*

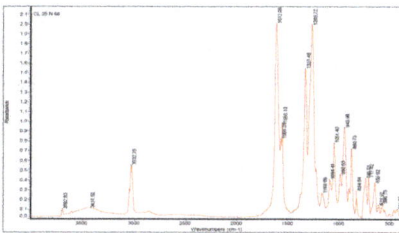

Figure 4. *FTIR Analysis*

Figure 5. *SEM microphotograph*

Table 1. *Experimental data of particle size*

Run Order	Experiment	Solution Pressure (bar)	Antisolvent Pressure (bar)	Antisolvent Temperature (°C)	Stirrer Speed (RPM)	Solution Concentration (%)	Nozzle Diameter (mm)	Particle Size (μm)
1	N-19	6	6	-9	800	5	0.7	5.63
2	N-20	7	7	-9	800	5	0.7	5.77
3	N-26A	5	1	30	1010	10	0.7	4.19
4	N-45	5	1	0	1010	10	0.7	2.38
6	N-33	6	1	30	500	5	1.0	10.34
7	N-77	7	1	20	1020	10	0.5	8.67
8	N-94	6	1	20	1020	10	0.4	6.48
10	N-26	2	3.5	30	600	10	0.7	50.3
11	N-29	4	1	45	500	10	0.7	7.62
12	N-50	2	1	0	1080	10	0.7	14.2
13	N-51	2	1	13	1080	10	0.7	5.89
14	N-24	4	1	-2	800	5	1.0	3.76
16	N-23	6	1	-2	800	5	1.0	4.35
17	N-25	6	1	30	800	5	1.0	8.93
18	N-32	1	1	35	500	10	1.0	22.5
19	N-34	2	1	9	500	10	1.0	9.31

20	N-35	4	1	9	500	10	1.0	8.1
22	N-31	2	1	35	500	10	1.0	4.74
23	N-36	1	1	11	500	10	0.55	5.72
25	N-28	1	1	45	1120	10	0.55	19.8
26	N-55	1	1	10	1120	10	0.55	6.92
27	N-62	6.2	1	20	1040	10	0.5	7.05
28	N-68	6.5	1	20	1700	10	0.5	8.52
29	N-66	7	1	20	1700	10	0.5	5.34
30	N-79	6	1	15	1800	10	0.5	5.15
31	N-80	6.2	1	15	1500	10	0.5	6.03
33	N-53	2	1	10	1120	10	0.7	24.0
34	N-82	6	1	15	1500	10	1.5	21.3
35	N-83	6	1	10	1100	10	0.5	4.51
36	N-87	6	1	0	1100	10	0.5	3.40
37	N-88	6	1	0	1100	10	0.5	4.67
39	N-93	6	1	19	1020	10	0.5	8.0
40	N-95	6	1	20	1080	10	0.4	5.77
41	N-97	6	1	15	1020	10	0.4	11.1
42	N-98	6	1	5	1040	10	0.4	16.9

Table 2. Experimental testing data

Run Order	Exper iment	Solutio n Pressur e (bar)	Antisolve nt Pressure (bar)	Antisolvent Temperature (°C)	Stirrer Speed (RPM)	Solution Concentration (%)	Nozzle Diamete r (mm)	Particle Size (μm)
5	N-30	1	1	14	500	10	0.55	5.31
9	N-44	5	1	0	1010	10	0.70	2.69
15	N-56	1	1	06	1120	10	0.55	2.74
21	N-60	6	1	25	1040	10	0.50	2.57
24	N-76	7	1	22	1020	10	0.50	4.18
32	N-52	3	1	28	1120	05	0.70	2.92
38	N-81	6	1	13	1200	10	0.50	1.82

Explosion Shock Waves and High Strain Rate Phenomena
Materials Research Proceedings **13** (2019) 121-127

Materials Research Forum LLC
https://doi.org/10.21741/9781644900338-21

ANFIS Model Development

ANFIS architecture based on the first-order Sugeno model is shown in Fig.7.ANFIS was implemented using Matlab (Version 8.1, 2013a, The Mathworks, Inc., USA). In particular, the Matlab Fuzzy Toolbox was adopted as the means for building ANFIS. The data matrix was structured in the form of seven columns, with the first six corresponding to the input variables and the seventh one to the output variable (response). Out of 49 data sets, 42 sets of data (Table 1) were used for training the ANFIS model and 07 dataset were used for testing the model (Table 2). After testing the testing data the experimental value and the ANFIS predicted value, Average Testing Error is 0.2431. After the model is trained and tested, the rule viewer is used for finding the ANFIS output for any instance of input by just entering them at the bottom of the window as shown in Figure 8. For an instance when the six inputs 1, 1, 6, 1120, 10, 0.55 and output particle size predicted by the ANFIS is 2.54 μm.

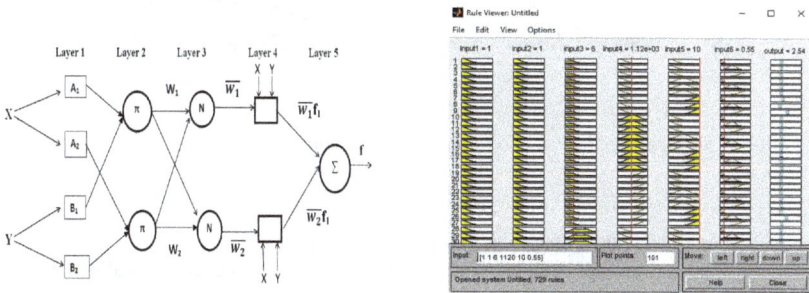

Figure 6. *Schematic of ANFIS architecture based and ANFIS rule viewer*

Comparison of ANFIS Predicted Value and Experimental Value of Particle Size

After training and testing the experimental data, the value of particle size, which is inbetween 1 to 6 μm, is compared with the experimental value and ANFIS prediction value is shown in Table 4. Actual error is the subtraction of experimental value and ANFIS prediction value. And Absolute Percentage Error is ratio of actual error to the experimental value as shown in Table 4.

Table 4. *Comparison of Predicted and Experimental Particle Size Results*

Run Order	Experiment	Nozzle Diameter (mm)	Particle Size (μm)		Error	
			Exp. Value	ANFIS Prediction	Actual Error (μm)	Absolute Percentage Error (%)
5	N-30	0.55	4.86	5.31	-0.45	09.25
9	N-44	0.70	2.69	2.38	0.31	11.52
15	N-56	0.55	2.74	2.54	0.20	07.29
21	N-60	0.50	2.57	2.81	-0.24	09.33
24	N-76	0.50	4.18	4.06	0.12	02.87
32	N-52	0.70	2.92	2.96	-0.04	01.36
38	N-81	0.50	1.82	1.77	0.05	02.74

Explosion Shock Waves and High Strain Rate Phenomena
Materials Research Proceedings **13** (2019) 121-127

Materials Research Forum LLC
https://doi.org/10.21741/9781644900338-21

Surface View of Particle Size

Apart from its prediction efficiency, ANFIS revealsinformation about the system being modeled. This is evidenced by the three-dimensional diagrams of the response variable as a function of the causal factors (in pairs of two), as presented in Figure 12. From this figure, it is apparent that nonlinear relationships between the input process variables and the response variable were well represented with response surface predicted by ANFIS. It appears that ANFIS captures such relationships and could be used as a tool where approximations of such relationships are required.

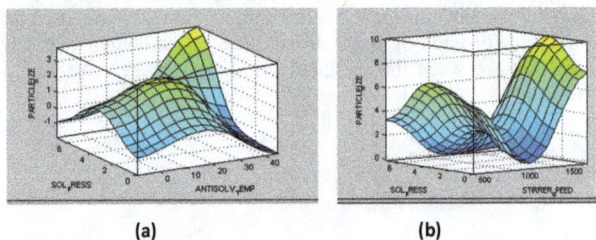

(a)   (b)

Figure 7. Surface response: (a) Solution pressure & Antisolvent temp (b) solution pressure & stirrer speed

Application of UF-CL20 Explosive in Slapper Detonator

In slapper detonator, prompt detonation of high density (1.5 g/cc) pellet of ultrafine CL-20 occurs when a thin polyimide flyer is impacted. The flyer is launched to a velocity of 2-3 km/s by electrically exploding foil when high electrical energy applied in short duration. It impacts the CL-20 explosive leading to initiation by shock-to-detonation phenomenon.Experimental prototypes of slapper detonators were fabricated and threshold trials were carried out using Bruceton test. The threshold initiation voltage for CL-20 is the range of 0.9 kV with standard deviation of ±0.1 kV. The threshold initiation energy for CL-20 based slapper detonator is much lower than HNS-IV based slapper detonator.

Conclusions

The main advantage of NASAS process is recrystallization of ultrafine CL-20 explosive using solvent-antisolvent method. There exists nonlinear relationship between the input process variables and the output variable 'particle size'. Neuro-fuzzy technique 'ANFIS'was developed for capturing such relationships and predicting the particle size for a given set of input variables. The ANFIS model was trained and tested using experimental data. Maximum absolute percentage error between the predicted value and experimental value is 11.52% which is in acceptable range. The model will further become accurate when large amout of training data is used. Artifical Neural based techniques enable learning and improving process characteristics for obtaining desired and repeatable output. It is experimentally found that the ultrafine CL-20 explosive based Slapper detonator has a threshold of 0.9 kV ± 0.1 kV. It is a type of low energy exploding foil initiator which has a great potential for futuristic applications in miniaturized systems such as Fuzes for various munitions and warheads.

Explosion Shock Waves and High Strain Rate Phenomena
Materials Research Proceedings **13** (2019) 121-127

Materials Research Forum LLC
https://doi.org/10.21741/9781644900338-21

Acknowledgements

The authors would like to gratefully thank Dr. Manjit Singh, Director TBRL for his support and encouragement. This research did not receive any specific grant from funding agencies in the public, commercial, or not-for-profit sectors.

References

[1] J. Wang, et al., Study on Ultrasound-and Spray-Assisted Precipitation of CL-20. Propellants, Explosives, Pyrotechnics, 37(6) (2012), 670-675. https://doi.org/10.1002/prep.201100088

[2] D. Spitzer, et al., Continuous crystallization of submicrometer energetic compounds, Propellants, Explosives, Pyrotechnics, 36(1) (2011), 65-74. https://doi.org/10.1002/prep.200900002

[3] M. Klaumünzer,J. Hübner, and D. Spitzer, Production of Energetic Nanomaterials by Spray Flash Evaporation, World Academy of Science, Engineering and Technology, International Journal of Chemical, Molecular, Nuclear, Materials and Metallurgical Engineering, 10(9) (2016), 1191-1195.

[4] G. Yang,et al., Preparation and Characterization of Nano-TATB Explosive, Propellants, Explosives, Pyrotechnics: An International Journal Dealing with Scientific and Technological Aspects of Energetic Materials, 31(5) (2006), 390-394. https://doi.org/10.1002/prep.200600053

[5] S. Gupta, D. K.Pal, et al., Pressurized Nozzle-Based Solvent/Anti-Solvent Process for Making Ultrafine ε-CL-20 Explosive, Propellants, Explosives, Pyrotechnics, 42(7) (2017), 773-783. https://doi.org/10.1002/prep.201700002

[6] S. Gupta, et al., D. K. Pal, et. al., Micro Nozzle Assisted Spraying Process for Re-crystallization of Submicrometer Hexanitrostilbene Explosive, Propellants, Explosives, Pyrotechnics, 43 (7) (2018), 721-731. https://doi.org/10.1002/prep.201800008

[7] S. Singh, et al., Neural network analysis of steel plate processing, Iron making and Steelmaking, 25(5) (1998), 355-365.

[8] J. M. Vitek,Neural networks applied to welding: two examples, ISIJ international, 39(10) (1999), 1088-1095. https://doi.org/10.2355/isijinternational.39.1088

[9] I.S. Kim, et al., Optimal design of neural networks for control in robotic arc welding, Robotics and computer-integrated manufacturing, 20(1) (2004), 57-63. https://doi.org/10.1016/s0736-5845(03)00068-1

[10] S. Pal, S.K. Pal, and A.K. Samantaray, Artificial neural network modeling of weld joint strength prediction of a pulsed metal inert gas welding process using arc signals, Journal of materials processing technology, 202(1-3) (2008), 464-474. https://doi.org/10.1016/j.jmatprotec.2007.09.039

[11] A.K Pannier,R.M. Brand, and D.D. Jones, Fuzzy modeling of skin permeability coefficients, Pharmaceutical research, 20(2) (2003.), 143-148.

[12] J. Jang,Neuro-fuzzy modeling: architectures, analyses and applications [dissertation]. California: University of Berkeley, 1992.

[13] J. Hines,L.H. Tsoukalas, and R.E. Uhrig, MATLAB supplement to fuzzy and neural approaches in engineering, John Wiley & Sons, Inc., 1997

[14] J. S. Jang, ANFIS: adaptive-network-based fuzzy inference system, IEEE transactions on systems, man, and cybernetics, 1993, 23(3), 665-685. https://doi.org/10.1109/21.256541

Explosion Shock Waves and High Strain Rate Phenomena   Materials Research Forum LLC
Materials Research Proceedings **13** (2019) 128-133   https://doi.org/10.21741/9781644900338-22

Explosive Welding of Al-MS Plates and its Interface Characterization

Bir Bahadur Sherpa[1,2,3,a*], Pal Dinesh Kumar[2,b], Abhishek Upadhyay[2,c]
Sandeep Kumar[2,d], Arun Aggarwal[2,e], Sachin Tyagi[1,3,f]

[1]Academy of Scientific and Innovative Research (AcSIR) Ghaziabad-201002, India

[2] DRDO-Terminal Ballistics Research Laboratory (TBRL), Sector-30, Chandigarh 160030, India

[3]CSIR-Central Scientific Instruments Organisation (CSIO), Sector-30, Chandigarh 160030, India

[a]sherpa7419@gmail.com, [b]dineshpalb@yahoo.com, [c]abhiups007@gmail.com,
[d] sandeepk44@gmail.com, [e]arun907@yahoo.com, [f]sachintyagi.iitr@gmail.com

*Corresponding author: sherpa7419@gmail.com Tel: +91-1733-305062, Fax: +91-172-2657506

Keywords: Explosive Welding, Impact Process, Hardness Value, Interface

Abstract: Explosive welding is a solid state welding process in which two similar or different materials are claded with the help of explosive energy. The high pressure generated during the process helps to achieve the interatomic metallurgical bonding in the two materials. In this research work, 5 mm aluminum plate was cladded with 20 mm mild steel for plate length of 300 mm x100 mm. Here parallel plate explosive welding set-up configuration using low VoD explosive consisting of mixture of Trimonite-1 and common salt was used. The interface joints were analyzed using optical inverted metallurgical microscope, SEM and Vickers Micro-hardness. It was observed that the value of micro-hardness at the interface was high as compared to the parent materials and decreased as we move away from the interface on both the sides. The optical and the SEM analysis showed straight morphology at most of the welded area. Al-MS plates were successfully welded using this low VoD explosive.

Introduction

Composite material with good corrosion resistant as well as bond strength is one of the prime needs of any industry for their respective work application. Explosive welding is a well known defined solid state weld process, where two plates are cladded with the help of explosive energy in which flyer plate is accelerated towards the base plate and at the interface a very high pressure order of magnitude 10^2 Mbar is generated followed by jet phenomenon[1]. Jet phenomenon is one of the important conditions for welding which occurs at the collision point in which it removes the oxide layer and provide clean mating surface free of contamination. This is attained by high pressure and kinetic energy deposited during the welding process[2]. Jet process helps atoms of two materials to meet at interatomic distance and form a strong metallurgical bond, where high temperature is obtained followed by rapid cooling in order of 10^5k/s[3]. Beside this, for weld to occur the pressure should be sufficient high and for sufficient length of time to achieve the bond formation. In explosive welding, pressure generated exceeds the yield strength of both the materials and which act as fluid at the collision point. It is a critical joining process where different parameters such as collision velocity, flyer plate velocity, VoD of explosive plays a very important role in formation of good bond[2] [4]. Many researchers have worked on this process using different material combination with variable explosive properties [5] [6] [7]. Aluminum is a light and corrosion resistant material having vast application in the naval and oil industries. The challenge of joining comes due to difference in chemical, physical properties as well as low solubility of iron in aluminum. Different means have been used to join this combination such as magnetic pressure

Explosion Shock Waves and High Strain Rate Phenomena   Materials Research Forum LLC
Materials Research Proceedings **13** (2019) 128-133   https://doi.org/10.21741/9781644900338-22

steam welding[8], diffusion bonding[9] ultrasonic welding[10] but all have some limitations and to circumvent this, explosive welding was adopted due to the advancement of joining any material combination with all size in a very few micro seconds. In this research work, aluminum is claded with mild steel and their behavior at interface is studied through micro-hardness & micro-structural analysis by preparing specimen with standard dimensions.

Material and methods

In this present experiment, aluminum was claded with mild steel having dimensions of 300x100x5mm & 300x100x20mm to produce single bimetallic materials. The systematic view of the explosive welding is shown in Fig. 1. Where the two plates kept parallel and separated by distance called stand-off. Low VoD explosive consisting of mixture of Trimonite-1 and common salt having VoD in the range of 1650-1800m/s was used. Aluminum as flyer plate was selected because of its light weight and superior corrosion resistant properties and mild steel as base plate due to its good tensile strength. The mechanical properties of the cladded material are shown in Table.1 and the experimental parameters used are shown in Table.2. In this experiment, low VoD explosive was used to provide necessary energy for metallurgical bonding using parallel set-up arrangement[11]. The practical set-up for explosive welding is shown in Fig.2 (A), where the impact on the ground and on the plate after the explosion is shown in Fig.2 (B), which shows how much energy is generated during the welding process. The bimetallic plate formed after the process is shown in Fig.2 (C), where bonded plates were then subjected to micro-hardness & micro structural examination, where in micro-hardness, the samples were prepared in the standard size and then examined the variation of hardness near the interface with the help of Vickers micro-hardness machine.

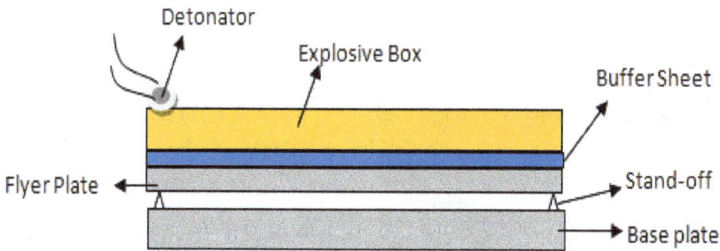

Fig.1. Systematic process diagram of explosive welding process

Explosion Shock Waves and High Strain Rate Phenomena   Materials Research Forum LLC
Materials Research Proceedings **13** (2019) 128-133   https://doi.org/10.21741/9781644900338-22

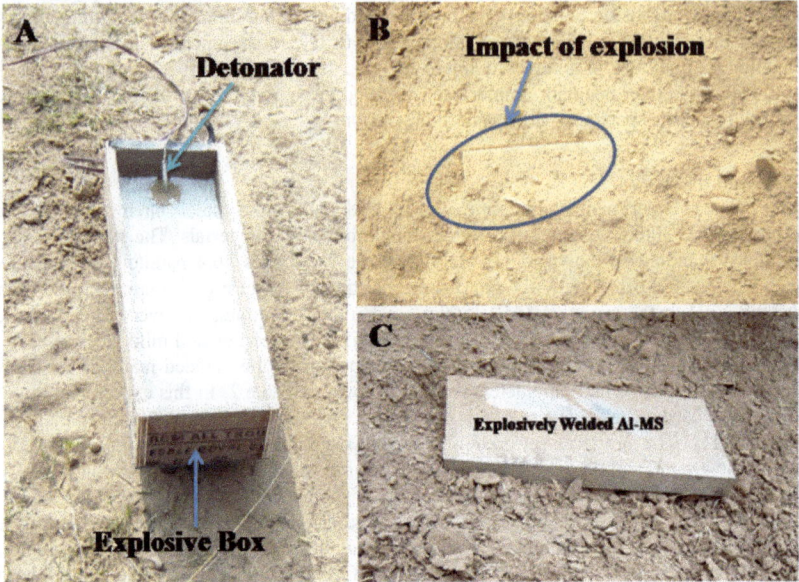

Fig.2. Overall set-up of explosive Welding A) Explosive box placed over plate B) Impact of explosion C) Bimetallic welded plates

Further for micro structural examination the samples were manually polished to 1μm finish. To study the internal structure, specimens were etched with 2% natal. This etched sample was analyzed in metallurgical optical microscope and to see the structural behavior near interface SEM was used. To check the transfer of elements during explosive welding dot mapping was performed. The successfully claded material was then subjected for further application.

Table.1 Mechanical Properties of Materials

Parameter	Aluminum (Al) Pure	Mild steel (MS) 1020
Hardness Value in (measured Vickers's)	41	160
Melting Point (Kelvin)	928	1789
Density (Kg/cm^3)	0.0027	0.00786
UTS (MPa)	90	340
Thermal conductivity (W/mK)	235	51.1
Specific Heat (J/Kg K)	904	486

Explosion Shock Waves and High Strain Rate Phenomena
Materials Research Proceedings **13** (2019) 128-133

Materials Research Forum LLC
https://doi.org/10.21741/9781644900338-22

Tabel.2 Experimental parameters

Parameter	Value
Flyer Plate with Dimension (mm)	Pure Aluminum (300x100x5)
Base Plate with Dimension (mm)	Mild Steel 1020(300x100x19)
Explosive	Trimonite with Salt Mixture
Loading Ratio	1.2
Stand-off Distance (mm)	6
Velocity of Detonation (m/s)	1550-1650

Results and Discussion

The bonded joints were examined by different equipment such as Vickers hardness, optical microscope, SEM. The morphology of the interface was straight with slight wavy at some interval as observed in both optical (Fig.3) and SEM images (Fig.4). Further to analyze the transfer of elements at the interface dot mapping (Fig.5) was done. Where, a sharp and clear image of elements at the interface confirmed good bonding. Micro-hardness indentations of first two points from either side along with interface point were plotted against hardness Vs distances are shown in Fig.6. The average hardness value of the mild steel was 157HV and that of aluminum was 41HV initially. It was observed that the hardness value at the interface had increased to 210HV as compared to adjacent material. The hardness value was increasing towards interface and stable after some distance away from interface. This similar increase in hardness was also reported by Mudali et al. [12] for titanium & stainless steel and by Abhishek et al. [6] for SS304 & AA6061.The increase in hardness is mainly due to grain dislocation pileup from shock hardening during explosive welding process & can also be observed in optical image as shown in Fig.3.Further the grain boundaries containing ferrite & pearlite were observed by adding etchant of Nital (2%).

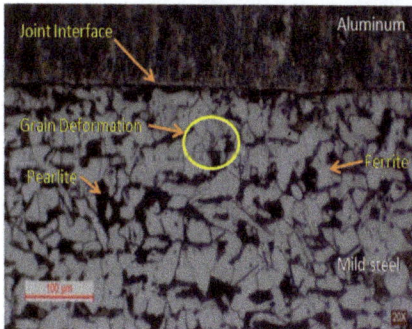

Fig.3. Optical structure of welded Al-MS plates   Fig.4. SEM image for bonded Al-MS interface

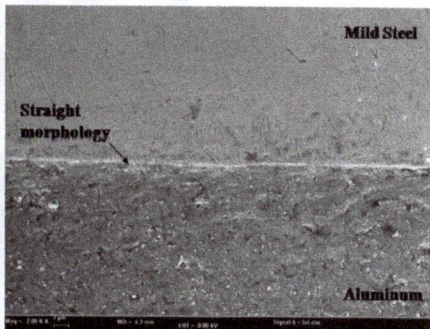

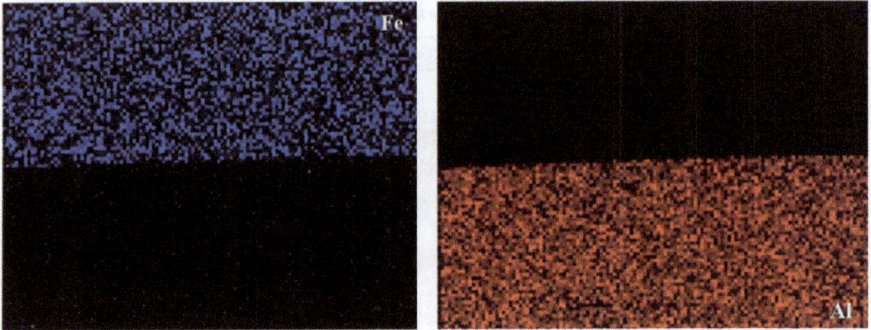

Fig.5. Dot mapping of the bonded Al-MS interface

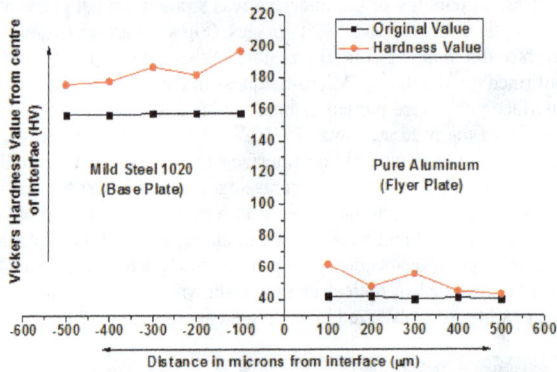

Fig.6. Micro-Hardness profile across interface of aluminum & mild steel

Conclusion

Pure aluminum and mild steel 1020 bimetallic plate were claded successfully. For joining these two incompatible materials, low velocity of detonation (VoD) explosive having VoD in range of 1650-1650 m/s was applied. Microhardness examination showed an increase in hardness value towards the interface as noticed from both the materials end. This increase in hardness can also be correlated with the grain deformation as observed in optical microscope images at the interface. In microstructural examination we observed straight morphology at the weld interface of the welded bimetallic plates.

Acknowledgements

The support from Terminal Ballistics Research Laboratory is highly acknowledged. Authors are very thankful to Dr. Manjit Singh, Director, TBRL Chandigarh and all the scientists, M.T section & workshop of TBRL for their valuable support.

Reference

1. Bernard Crossland, *Oxford Series of Advanced Manufacturing 2* 1982, p. 233.

Materials Research Forum LLC
https://doi.org/10.21741/9781644900338-22

2. AA Akbari Mousavi and STS Al-Hassani, *Journal of the Mechanics and Physics of Solids* 2005, vol. 53, pp. 2501-2528.

3. Tadeusz Zdzislaw Blazynski: *Explosive welding, forming and compaction.* (Springer Science & Business Media, 2012).

4. Bir Bahadur Sherpa, Pal Dinesh Kumar, Abishek Upadhyay, Uma Batra and Arun Agarwal, *Advances in Applied Physical and Chemical Sciences-A Sustainable Approach* 2014, pp. 33-39.

5. K Raghukandan, *Journal of Materials Processing Technology* 2003, vol. 139, pp. 573-577.

6. Abhishek Upadhyay, Bir Bahadur Sherpa, Sandeep Kumar, Niraj Srivastav, Pal Dinesh Kumar and Arun Agarwal, In *Materials Science Forum,* (Trans Tech Publ: 2015), pp 261-264. https://doi.org/10.4028/www.scientific.net/msf.830-831.261

7. SAA Akbari Mousavi and P Farhadi Sartangi, *Materials & Design* 2009, vol. 30, pp. 459-468.

8. Kwang-Jin Lee, Shinji Kumai, Takashi Arai and Tomokatsu Aizawa, *Materials Science and Engineering: A* 2007, vol. 471, pp. 95-101.

9. PS Gawde, R Kishore, AL Pappachan, GB Kale and GK Dey, *Transactions of the Indian Institute of Metals* 2010, vol. 63, pp. 853-857. https://doi.org/10.1007/s12666-010-0130-x

10. N Mohan Raj, LA Kumaraswamidhas, Pavan Kumar Nalajam and S Arungalai Vendan, *Transactions of the Indian Institute of Metals* 2018, vol. 71, pp. 107-116. https://doi.org/10.1007/s12666-017-1140-8

11. Bir Bahadur Sherpa, Abhishek Upadhyay, Sandeep Kumar, Pal Dinesh Kumar and Arun Agarwal, *Materials Today: Proceedings* 2017, vol. 4, pp. 1260-1267.

12. U Kamachi Mudali, BM Ananda Rao, K Shanmugam, R Natarajan and Baldev Raj, *Journal of Nuclear Materials* 2003, vol. 321, pp. 40-48.

Explosion Shock Waves and High Strain Rate Phenomena
Materials Research Proceedings 13 (2019) 134-140

Materials Research Forum LLC
https://doi.org/10.21741/9781644900338-23

Effect of Liner Cone Angle, Liner Thickness and Wave Shaper in Large Caliber Shaped Charge Warheads

Mukesh Kumar*, Yashpal Singh, Pravendra Kumar

Scientist, Terminal Ballistics Research Laboratory, Sector -30 Chandigarh-160030, India

*E mail:mukesh98519@gmail.com

Keywords: Shaped Charge Warheads, Hydrocode, Wave Shaper, Autodyn 2D

Abstract. Shaped charge warheads are being utilized in defence applications against a wide variety of targets provided by armour, RCC and soil cover. Shaped charge warhead focus the explosive energy by the use of a cavity lined with metal normally called a liner. The concentration of energy along the axis of the warhead acts as force multiplier and hence lighter warheads are possible for deeper penetration. Performance of the shaped charge warhead is function of jet tip velocity, jet length and break up time (BUT). These performance parameters are greatly influenced by liner geometry, liner thickness and liner cone angle and selection of explosive. In this paper, simulations using AUTODYN numerical hydrocode were carried out to study the effect of liner geometry (Tulip vs conical), liner cone angle $(50^0,60^0,70^0,80^0)$ and liner thickness(4mm,6mm,8mm,10mm and 12mm) on large caliber shaped charge warheads. Numerical simulations were also done to study the effect of wave shaper in shaped charge warhead. A shaped charge warhead of dia.340mm has been designed by using AUTODYN numerical hydrocode. OFE Copper (ASTM B152 C10100) is used as liner material. A wave shaper of dia.210mm and nylon material was used in shaped charge warhead. An Eulerian approach was used for the liner, casing, wave shaper and explosive parts. A single point initiation in the centre of the rear end of warhead was chosen. The numerical simulation results showed that the jet- tip velocity decreases in between 15-20% of liner position with increasing the cone angle when the other parameters are the same. For the cone angle 60^0, jet tip-velocity decreases as liner thickness is increased from 4mm ($V_{j\text{-tip}}$: 8.14 km/s) to 12mm ($V_{j\text{-tip}}$: 6.7 km/s). It was also realized that in case of wave shaper warhead there is more than 15% increase in jet tip velocity and 10% increase in jet length in comparison to without wave shaper warhead due to increase in collapse velocity of liner elements. The slug velocity is 1.22km/s in case of with wave shaper warhead whereas it was 1.05 km/s in without wave shaper. It means that a decision for the selection of liner geometry and dimensions of a shaped charge penetrator should be done according to target, required desired effect on target, permissible weight and available space for the warhead.

1. Introduction

The high-speed jet resulting from the detonation of shaped charge is used to penetrate and demolish hardened targets in military and commercial area. Particularly, the shaped charge is mainly used to penerate armoured vehicles such as tanks, to demolish bunkers, fuel tanks and bridges constructions. In addition , warheads designed by using the shaped charge principle are also used against submarines targets. In the industry, shaped charge are generally used in geophysical areas e.g. petroleum research, mining, steel industry and well bore penetration, underwater trenching and demolition. Two- and Three-dimensional numerical simulation of shaped charges play an important role in testing and designing of shaped charges. In recent studies,computer programs based on Eulerian and Lagrangian calcultions such as "Hydrocode"are utilized to design shaped charges warheads. Generally, the programs provide a good tool facility to analyze problems under very large material deformations.

Explosion Shock Waves and High Strain Rate Phenomena   Materials Research Forum LLC
Materials Research Proceedings **13** (2019) 134-140   https://doi.org/10.21741/9781644900338-23

In this paper, a shaped charge warhead of dia.340mm has been designed using AUTODYN numerical hydrocode simulation using 2D-axisymmetric model. The effect of liner geometry, liner cone angle, explosive and liner thickness were studied on large caliber shaped charge warhead. Jet characteristics such as jet tip velocity, slug velocity, BUT and jet length of shaped charge warheads with wave shaper and without wave shaper were also calculated. OFE Copper (ASTM B152 C10100) is used as liner whereas Comp B is used as explosive. A wave shaper of dia.210mm with nylon material is used in shaped charge warhead.

2. Description of shaped charges

The shaped charge principles were first used in developing weapons by Germans during World War-I. At present, the geometry of the modern shaped charges consists of a cylinder of explosive with a hollow cavity in one end and a detonator at the other end. The hollow cavity is usually lined with a thin layer of metal, glass and ceramics. The cross section of a typical shaped charge is given in Fig.1.

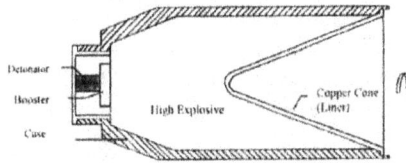

Fig.1- Cross-section of a typical shaped charge

The ignition of detonator causes a shock wave, which detonates the high explosive. The detonation waves which propagate spherically into the explosive moves at a very large velocity, around 5-10km/s.When the detonation wave reaches the conical liner surface, a conical liner is accelearted from apex to the base on the cylinder axis under the high detonation pressure,collapisng the liner . As the conical liner material is loaded with high energy, it collapses on the axis, which generates a hypervelocity jet. It is commonly known that the inner liner material on the cavity side forms the jet-tip which has an extremely high velocity, while the outer of the liner material which is in the contact with high explosive forms a jet tail called "slug" which is massive but has slow velocity.The front of the jet (jet-tip) has a velocity range 5-10km/s , while the back of the jet (jet-tail) has a velocity range of 0.5-1.0km/s The conical liner collapse and jet formation process in a shaped charge after initiation of detonation is shown in Fig.2

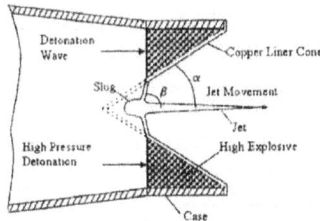

Fig.2 – Liner collapse and jet formation process

3.   Numerical simulation

3.1 Finite element analaysis:

In this section, the finite–element model will be described. The following section contains the analysis of the jet formation.

3.2 Simulation tools used:

ANSYS AUTODYN 2D/3D non-linear hydrocode was used. It is an explicit numerical analysis code, where the equations of mass, momentum and energy conservation coupled with materials descriptions are solved. Alternative numerical processors are available and can be selectively used to model different regions of a problem. The currently available processors include Lagrange, Euler, Euler FCT, ALE (Arbitrary Lagrange Euler) and SPH (Smooth particle Hydrodynamics), shell which make them suited to a wide range of non-linear dynamics problem. The codes are particularly suited to the modeling of impact, penetration, blast and explosive events.

3.3 Structure of shaped charge :

On the basis of theoretical analyis,a shaped charge of dia.340mm and warhead length 450mm has been designed having length/diameter ratio 1.3 by AUTODYN2D software. Liner is having a thickness of 4mm with a cone angle of 60^0.OFE Copper (ASTM B152 C10100) is used as liner whereas Comp B is used as explosive. Comp-B which is a secondary high explosive type is selected as high explosive material in the simulations. It is a mixture of 60% RDX and 40% TNT by weight.

(A) Effect of Wave shaper in large caliber shaped charge warhead:

Numerical simulations were carried out to study the effect of wave shaper in shaped charge warhead. Jet characterisctics such as jet tip velocity,slug velocity and jet length of shaped charge warheads with wave shaper and without wave shaper were calculated at time t=350μs.

(i) Numerical simulation results without wave shaper shaped charge warhead:

Jet-tip velocity and slug velocity with and without wave shaper in large caliber shaped charge warhead are shown in Table: 1.

Table 1: Jet- tip velocity with and without wave shaper

S.No.	Liner Thickness	Without wave shaper Jet- tip velocity(km/s)	With wave shaper Jet tip velocity(km/s)
1.	4mm	6.08	8.14
2.	6mm	5.88	7.60
3.	8mm	5.59	7.20
4.	10mm	5.02	6.95

Materials Research Forum LLC
https://doi.org/10.21741/9781644900338-23

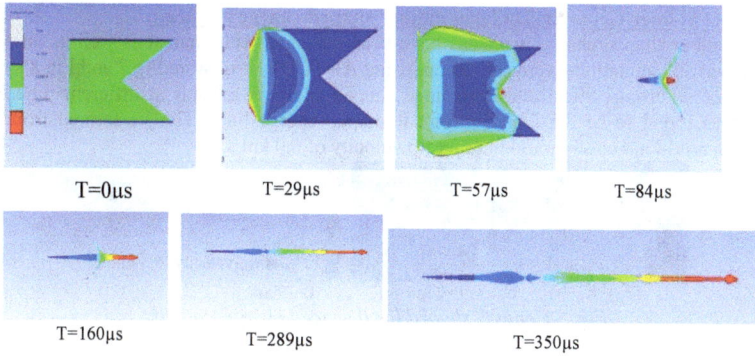

Fig. 5 Formation of jet without wave shaper at different time steps

(ii). Numerical simulation results with wave shaper shaped charge warhead:

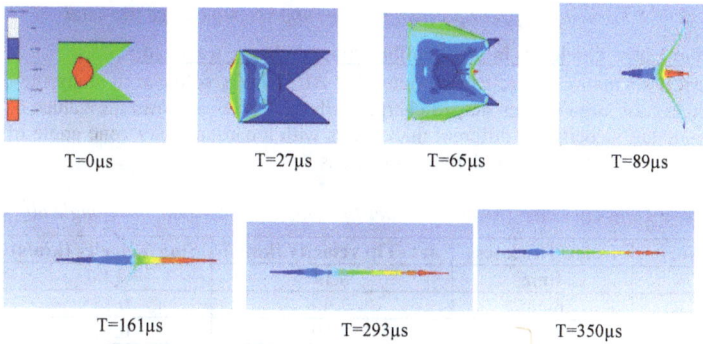

Fig. 6 Formation of jet with wave shaper at different time steps

Fig.5 and Fig.6 shows jet formation at different time steps without and with wave shaper shaped charge warheads.

Table- 2 Comparison of different parameters between without and with wave shaper SC warheads

Parameters	Without wave shaper SC warhead	With wave shaper SC warhead
Length of Jet(L)(in mm)	1490	1644
Jet-tip velocity (Vj-tip)(in km/s)	6.08	8.14
Slug Velocity(in km/s)	1.05	1.22

Explosion Shock Waves and High Strain Rate Phenomena
Materials Research Proceedings **13** (2019) 134-140

Materials Research Forum LLC
https://doi.org/10.21741/9781644900338-23

(B) Effect of liner Geometry in large caliber shaped charge warhead:

Large caliber shaped charge warheads have been designed with conical as well as tulip shaped liner. Numerical simulations were carried out using AUTODYN software.Fig.7 and Fig.8 shows the formation of jet from conical and tulip lined shaped charge warheads at different times. Jet tip velocity was found to be 8.5 km/s in conical shaped charge liner. The jet formed from shaped charge with tulip liner configuration has jet tip velocity of 6.0 km/s.

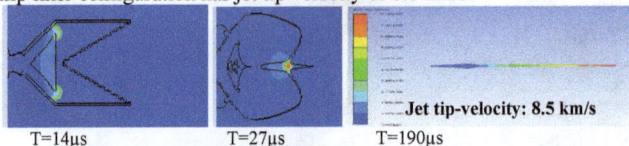

T=14µs   T=27µs   T=190µs   **Jet tip-velocity: 8.5 km/s**

Fig. 7 Conical shaped lined shaped charge warhead

Jet tip-velocity: 6.0 km/s

Fig. 8 Formation of Jet-tip velocity from Tulip shaped charge warhead

(C) Effect of Liner Thickness in large caliber shaped charge warhead:

Numerical simulations were carried out using AUTODYN software to study the effect of liner thickness for large caliber shaped charge warhead. Table- 3 shows the values of jet –tip velocity and slug velocity for different thicknesses with a constant liner cone angle of 60^0 . A nylon wave shaped was also used during simulations.

Table -3 Different Liner Thickness vs. Jet-tip velocity for constant cone angle 60^0

S.No.	Liner Thickness	Jet -Tip velocity (km/s)	Slug Velocity (km/s)
1.	4mm	8.14	1.06
2.	6mm	7.60	0.785
3.	8mm	7.19	0.671
4.	10mm	6.95	0.522
5.	12mm	6.70	0.517

(D) Effect of liner cone angle:

2D simulations were carried out to study the effect of liner cone angle for constant liner thickness 4mm. Table- 4 shows the values of jet –tip velocity and slug velocity at different liner cone angles.

Table -4 Different Liner cone angle vs. Jet-tip velocity for constant liner thickness 4mm

S.No.	Liner cone angle	Jet -Tip velocity (km/s)	Slug Velocity (km/s)
1.	50^0	8.55	1.80
2.	60^0	8.14	1.06
3.	70^0	7.63	1.34
4.	80^0	7.58	1.53

Explosion Shock Waves and High Strain Rate Phenomena
Materials Research Proceedings 13 (2019) 134-140

Materials Research Forum LLC
https://doi.org/10.21741/9781644900338-23

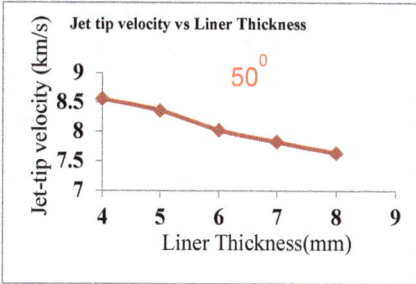

Fig. 9 Jet tip velocity vs. liner thickness for cone angle 50°

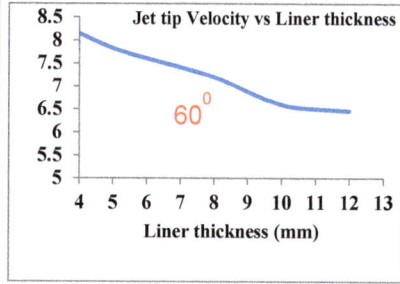

Fig. 10 Jet tip velocity vs. liner thickness for cone angle 60°

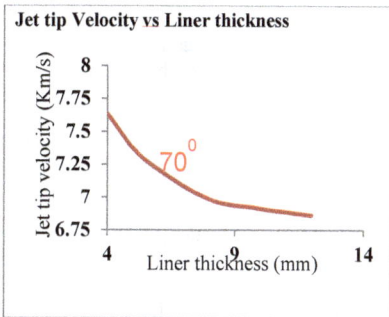

Fig. 11 Jet tip velocity vs. liner thickness for cone angle 70°

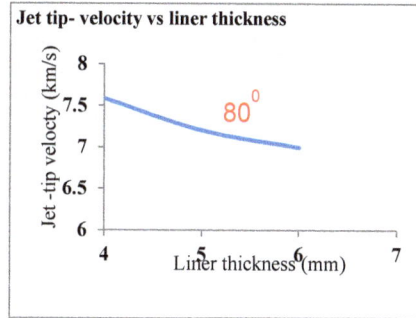

Fig. 12 Jet tip velocity vs. liner thickness for cone angle 80°

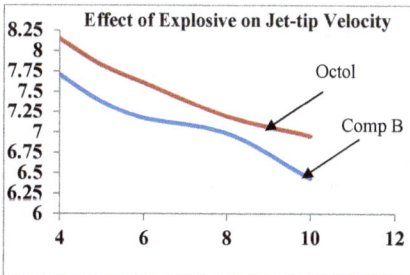

Fig.13 Effect of Explosive

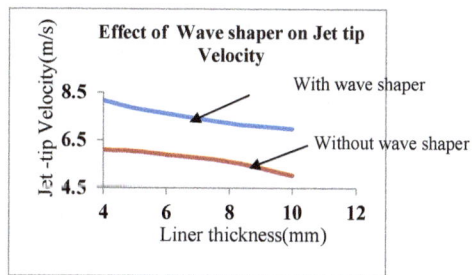

Fig.14 Jet tip velocity vs. liner thickness for cone angle 60°

4. Results and discussions

Table-1, 2, 3 and 4 show the results obtained from AUTODYN-2D numerical software. From simulations, it has been observed that jet-tip velocity is 6.08km/s without wave shaper based shaped charge warhead whereas it was 8.14km/s in case of with wave shaper warhead. Jet length is also 1490mm in case of without wave shaper warhead whereas it was 1644mm in case of with wave shaper based shaped charge warhead. It has been observed that in case of wave shaper warhead there is more than 15% increase in jet tip velocity and 10% increase in jet length in comparison to without wave shaper warhead due to increase in collapse velocity of liner elements. Jet tip-velocity is more in case of conical shaped charge liner in comparison to tulip shaped charge liner geometry but jet diameter is more in tulip liner geometry. The numerical simulation results showed that the jet tip velocity decreases in between 15-20% of liner position with increasing the cone angle when the other parameters are the same. In the simulations performed for different cone angles $50^0, 60^0, 70^0, 80^0$ it is observed that as an increase of cone angles, the jet-tip velocity decreases and slug velocity increases. In addition, as an increase of cone angle, the forward jet-tip diameter increases. For the constant cone angle 60^0, jet tip-velocity decreases as liner thickness is increased from 4mm (V_{jtip}: 8.14 km/s) to 12mm (V_{jtip} : 6.70 km/s).It was also observed from simulations that jet-tip velocity was increased in between 8-10% when Octol was chosen in place of CompB as explosive.

Conclusion

The purpose of this study is to perform an optimum modelling on order to benefit the shaped charge in an effective manner. Validations of these codes have been done only by matching the experimental results of the shaped charge warheads. There is an appreciable increase in jet-tip velocity and jet length in case of wave shaper shaped charge warhead thus there is an increase in penetration in the target. It can also be concluded that large caliber shaped charge performance are greatly influenced by the line geometry, liner thickness, liner cone angle, explosive and wave shaper and a decision for the selection of liner geometry and dimensions of shaped charge penetrator should be done according to target, required desired effect on target, permissible weight and available space for the warhead.

References

1. W.P Walter, J.A Zukas "Fundamental of Shaped Charges" John Wiley Interscience Publication.
2. Theory Manual ANSYS AUTODYN
3. M.A Meyers, Dynamics Behavior of Materials, Wiley, New York, 1994
4. Mayseles M, Hirsch E, Lindenfeld A, "Effect of explosive in the shaped-charge jet formation characteristics", 16th Int. Symp. Ballistics, San-Francisco, USA, Sept. 23-28, 1996
5. J.F Molinari "Finite element simulation of shaped charges", Finite Elements in Analysis and Design, Vol.38, 2002, pp 381-389. https://doi.org/10.1016/s0168-874x(02)00085-9
6. I. Gokhan Aksoy , Sadri Sen "Effect of the variation of conical liner apex angle and explosive ignition point on shaped charge jet formation ", Indian Journal of Engg. & Material Sciences, Vol.10, Oct.2003, pp 381-389
7. Chen CY *et al.* "Design of an inert material type plane wave generator" Propellants Explosive, Pyro techniques, Vol.18, 1993, pp 139-145. https://doi.org/10.1002/prep.19930180305
8. G. Pezzica et al "Calculation of the wave –shaper effects on Detonation wave in shaped charges" Propellants, Explosive, Pyrotechnics, Vol. 12, 1987, pp 125-129. https://doi.org/10.1002/prep.19930180305

Explosion Shock Waves and High Strain Rate Phenomena
Materials Research Proceedings **13** (2019) 141-148

Materials Research Forum LLC
https://doi.org/10.21741/9781644900338-24

Improvement in Performance of Shaped Charge using Bimetallic Liner

Santosh N. Ingole[1,a,*], M.J. Rathod[2,b], K.M. Rajan[3,c], R.K. Sinha[4d], S.K. Nayak[3,e], Nair Prakash N.P.[3, f], V.K. Dixit[3,g], S.G. Kulkarni[5,h]

[1]Directorate General Aeronautical Quality Assurance, Ministry of Defence, New Delhi- 110011, India

[2]College of Engineering Pune, Shivajinagar, Pune- 411005, India

[3]Defence Research and Development Organisation, Armament Research and Development Establishment Pashan, Pune- 411021, India

[4]Defence Research and Development Organisation, High Energy Materials Research Laboratory Sutarwadi, Pune- 411021, India

[5]Defence Intitute of Advanced Technology Girinagar, Pune – 411025, India

[*a]ssingole@rediffmail.com, [b]mjr.meta@coep.ac.in, [c]director@arde.drdo.in,[d]rksinha@hemrl.drdo.in,[e]sknayak@arde.drdo.in,[f]nairprakash@arde.drdo.in, [g]vkdixit05@rediffmail.com,[h]sgkulkarni@diat.ac.in

Keywords: Bimetallic Liner, Monolithic Liner, Penetration Depth, Tip Velocity, Slug Velocity

Abstract. Shaped charge has been designed by replacing conventional monolithic liner with bimetallic liner to possibly enhance its penetration capability. A shaped charge with bimetallic liner formulated using aluminum as the outer cone and copper as the inner cone with cone angle of 52^0,liner thickness of 2.4mm and calibre of 60mm. Theoretically predicted performance parameters have been compared with that of experimentally determined such as penetration depth and jet tip / slug velocity. The experimental results are reasonably in good agreement with theoretically predicted values. Penetration depth and tip velocity exhibit 22.5% and 10.67% increase respectively in comparison with shaped charge using copper as monolithic liner.

Introduction

Metallic lined shaped charge is frequently used in anti-armour projectiles for their higher penetration capabilities due to high velocity homogenous and unbroken jet formed after collapse of metallic liner by explosive action on the conical cavity of the projectile. The jet velocity is dependent upon several factors such as explosive used in the projectile, liner material, its density and liner thickness, cone angle, base of conical cavity, L/D ratio of the projectile, etc. [1].

Penetrative performance depends on cone angle as well. Smaller the cone angle ($\sim40^0$), narrower will be the jet and deeper will be the penetration. However, for higher lethality cone angle can be increased up to 120^0. Most of the High Explosive Anti-tank (HEAT) weapons presently being used are based on monolithic liner [2,3].

Attempts have been made by various researchers to improve the penetrative performance of shaped charge weapon by modifying the existing configuration. Replacement of monolithic liner with bimetallic liner is one of the preferred options for enhancement of performance of such weapon. Number of workers have reported bimetallic liner based shaped charge weapons using high density and high valued metals such as Au, Pt, Rh, Ta etc.[4]. Skolnick and Goodman [5] have elaborately discussed specific advantages for using multi-layered liners in combination with

Explosion Shock Waves and High Strain Rate Phenomena   Materials Research Forum LLC
Materials Research Proceedings **13** (2019) 141-148   https://doi.org/10.21741/9781644900338-24

different explosives in regards to enhancement of performance and also offered possible mechanism.

Chanteret et al [6] reported combination of two different liner materials, one with high sound velocity and other with high ductility, in designing the shaped charge. According to these authors, high tip velocity is controlled by the material having higher sonic velocity and supported by the material having high ductility.

In the present study, attempts have been made to improve penetrative performance by replacing monolithic liner with bimetallic liner consisting of aluminum (outer cone) and copper (inner cone). Theoretical prediction of the performance parameters for configured shaped charge has been carried out by computer simulation using hydrocode (AutoDYN) [7]. The theoretically predicted parameters have been experimentally validated and results are compared.

Experimental Section

Materials and methods
Materials and configuration of bimetallic liner shaped charge
The configuration details of bimetallic liner shaped charge (BMLSC) are summarized in Table 1:

Table 1: Configuration of BMLSC.

Nomenclature	Particulars	Remarks
Calibre	60mm	
Length	115mm	L/D~2
Main Charge	HMX-Wax	ρ= 1.758 g/cm^3, 380g
Booster	RDX-Wax	ρ=1.685 g/cm^3,10g
Liner Material (L1)	Copper	1.2 mm flow formed 52^0, 54mm
Liner Material (L2)	Aluminum	1.2 mm flow formed 52^0 , 54mm long
Case Material for shaped charge	Aluminum alloy	Material as per HE-30 standard

General configuration of typical BMLSC is given in figure 1.

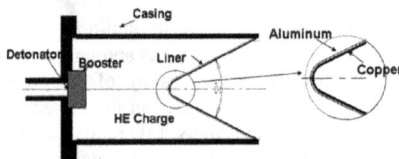

Figure 1. General configuration of BMLSC.

Methods

(A) Theoretical prediction of performance parameters by computer simulation.

BMLSC was modelled using nonlinear hydrocode-AutoDYN [8] assuming axial symmetry as it reduces computational time. The Aluminum and copper were used in different weight proportions (Al:Cu::45:55; 50:50; 55-45; 85:15) keeping the effective liner thickness as 2.4mm. Similar studies were also carried out for monolithic liner shaped charge (MLSC) consisting of copper as liner material for comparison. The penetration depth was predicted with mild steel target plates to

Materials Research Forum LLC
https://doi.org/10.21741/9781644900338-24

understand the penetration behaviour of BMLSC. The simulation results include the penetration depth and tip velocity for MLSC and BMLSC.

(B) Experimental evaluation

MLSC (Cu) and BMLSC (Al:Cu::50:50) have been fabricated as per the details included under methods. They were subjected to penetration and soft recovery trials. Jet parameters were determined by flash radiographic technique. Details of each experimental technique are included in the following sub-sections.

Penetration trial

Penetration depth of shaped charge was measured at standoff distance of 5 calibre in mild steel target plates stacked in vertical orientation to ascertain the penetration behaviour of BMLSC.. The test setup for penetration trial is given in Figure 2. Prediction of penetration depth was made under identical condition as in computer simulation.

Soft recovery trial

Recovery of shaped charge jet and slug particles were done in the experimental setup as described by Lassila et al [9]. Figure 3 includes the test setup used in the trial.

Various materials with increasing density such as air, aqueous film forming foam (AFFF), polystyrene (PS or Thermocole), polyethylene foam (PEF); polyurethane foam (PUF); high density polyurethane (HDPUF) and water were used in the test setup to capture jet/slug particles. The recovered jet and slug particles were subjected to metallographic investigations.

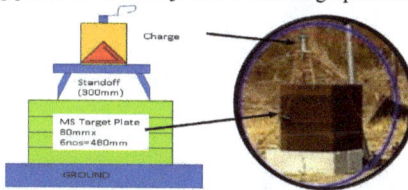

Figure 2. Test setup for penetration trials

Flash radiography(FXR)

FXR technique was used to record liner collapse of cone, jet formation, jet penetration, and slug formation. Synchronization pattern / schematic setup is depicted in figure 4. The velocity, dimension and breakup time of jet were calculated from FXR records. The FXR records experimentally obtained and predicted by computer simulation are included in figure 5.

Results and Discussion

Mechanism of jet formation in BMLSC

Figure 6 shows the collapse mechanism of liner and jet formation at different time intervals assuming the complete collapse and formation of fully emerged jet after several microseconds.

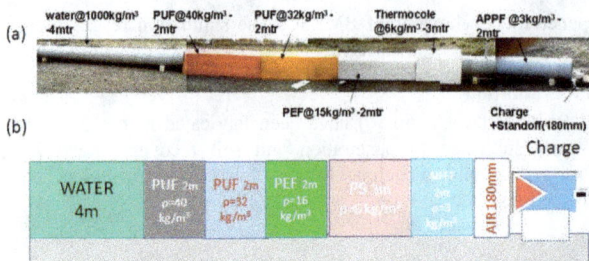

Figure 3. (a) Experimental set up and   (b) Plan for soft recovery trial.

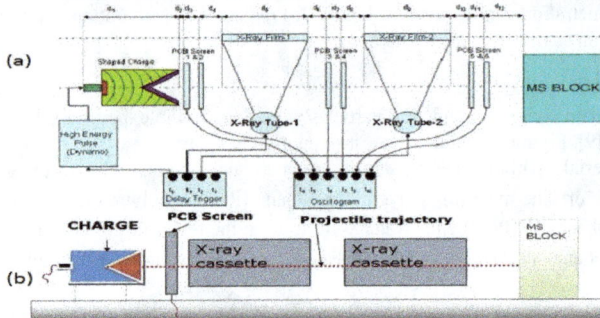

Figure 4. (a) Synchronisation plan; (b) Schematic setup for flash radiography.

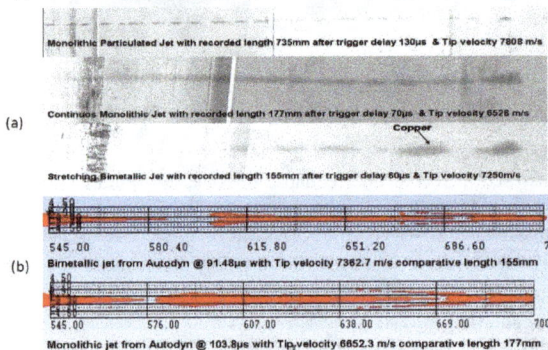

Figure 5. (a) FXR of jets; (b) comparative jet from computer simulation.

From figure 6, it is clear that for 50:50 BMLSC configuration, maximum portion of liner of inner cone constitutes the jet whereas maximum portion of liner of outer cone constitutes the slug. It is evident that depth of penetration is directly proportional to length of unbroken jet and under root of density of jet material and inversely proportional to under root density of target material. Thus, for better penetrative performance high density jet material is preferred. This is applicable to MLSC with copper as the liner material. In BMLSC, the low-density material like aluminum remains in slug and subsequently vaporizes out. Copper has higher density (8.9 g/cm^3) and melting point (1083^{0}C) than aluminum (density:2.71g/cm^3 and melting point: 657^{0}C) is preferred material for shaped charge application [10]. The sonic velocity of aluminum (6.4 km/s) is, however, higher than copper (4.7 km/s). In 50:50 configuration of BMLSC, the high-density material like copper constitutes only 50% of mass, but still exhibits improved penetrative performance mostly because aluminum accelerates the jet formation process. Jet contains more quantity of copper than aluminum and helps to get improved performance as per conservation of energy and momentum principles. The inner part of the bimetallic cone (which is not in contact with explosive) forms jet, which is squeezed out from the apex of inner cone while outer cone (in contact with explosive) moves slower with respect to the inner cone and forms residue in the form of slug. Thus, the metal in the conical liner divides in two parts of velocity gradient with reference to the location. Simulation results from the extant work confirm this hypothesis and justifies the use of bimetallic liner for improvement in penetrative performance.

Penetration trial

Table 2 includes the results of penetration trials.

Table 2. Depth of penetration determined from experiments

Charge	Configuration	Penetration(mm)	Remarks
A-5	Monolithic (Cu)	190	BMLSC exhibits improved penetration depth over MLSC
A-1	BiMetallic (Al:Cu::50:50)	220	

Soft recovery trials

Jet and slug particles recovered during the trials were subjected to metallographic investigation. Figure 7 shows the jet and slug material recovered during the experimental trial along with that obtained from computer simulation. From this figure; cut section of the slug, it is clear that outer cone encases the inner one. It is in line with the prediction made in computer simulation studies. Further, slug recovered from bimetallic liner has mass 19.25g against total liner mass of 94.10g which implies that the slug is only 20.45% of total liner mass and Al has been vaporized out with visible thin layer of Al-Cu alloy (duralumin) at certain places. Metallography of recovered slug reveals recrystallized grain structure, due to melting and realigning in the direction of flow of jet. Cut section of slug from BMLSC do not show any trace of Al or alloy of Al-Cu formed inside the core, thus indicating the vaporization of aluminum due to its lower melting point. However, outer portion of the slug exhibits formation of Al-Cu alloy on the surface due to interaction of both the metals at high temperature. Hardness of both the recovered slugs of BMLSC and MLSC remains in the same range viz.,48-65 BHN in monolithic and 54-58 BHN in bimetallic.

Flash radiography (FXR)

Results from FXR records are presented in Table 3. Exposed jet is shown in figure 5 with comparative picture from computer simulation.

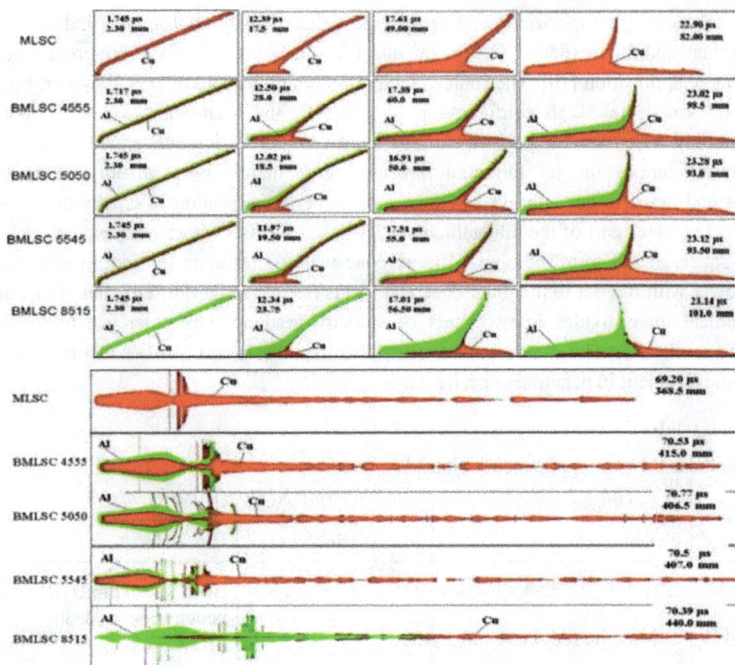

Figure 6. Liner collapse at various time intervals for different configurations.

Table 3: Results from flash radiography.

Charge	Configuration	Parameters	Observations
A-2	Monolithic(Cu)	Triggering delay T1-130µs & T2-300µs	Jet length recorded is 674mm and jet tip velocity at trigger delay of 130µs is 7832m/s
A-7	Monolithic(Cu)	Triggering delay T1-70µs & T2-630µs	Jet length recorded is 177mm and jet tip velocity at trigger delay of 70µs is 6528m/s
A-3	BiMetallic (Al:Cu::50:50)	Triggering delay T1-60µs & T2-430µs	Jet length recorded is 155mm and jet tip velocity at trigger delay of 60µs is 7250m/s

The results of penetrative performance based on computer simulation studies and experimental evaluation on are presented in Table 4. The error in computer simulation owes due to material characteristics. However, all the results are comparable from BMLSC and MLSC tested against standard conditions.

Materials Research Forum LLC

https://doi.org/10.21741/9781644900338-24

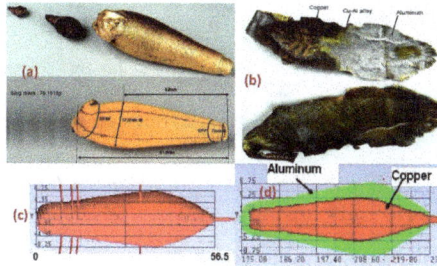

Figure 7. (a) MLSC jet/slug particle ; (b) BMLSC slug; (c) slug for MLSC from computer simulation and (d) slug for BMLSC from computer simulation

Table 4: Penetrative performance results based on computer simulation studies and experimental evaluation.

Configuration	Result Comparison			
	Performance Parameter	Computer simulation	Experimental evaluation	Remarks (error)
MLSC (Cu)	Penetration depth	310mm	190mm	64%
	Tip velocity at 70 µs	6652m/s	6528m/s	2%
	Slug mass to original mass %	-	58.85%	-
BMLSC (Al:Cu::50:50)	Penetration depth	380mm	220mm	72%
	Tip velocity at 60 µs	7362m/s	7250m/s	1.5%
	Slug mass to original mass %	-	20.45%	-
MLSC vs BMLSC	Penetration depth	Increaed by 22.58%	Increaed by 15.79%	Difference is comparable
	Tip velocity at 70 µs	Increased by 10.67%	Increased by 11.06%	
	Slug mass to original mass %	-	Decreased by 65.25%	Substantial difference

Conclusion

Based on the results obtained from the computer simulation studies and limited experimental trials, it can be concluded that bimetallic liner shaped charge exhibits better penetrative performance than monolithic liner shaped charge.

It was also realized from the studies, that the liner of the outer cone (aluminium) in BMLSC configuration mostly goes in the slug, a large portion of it vaporizes out, whereas, liner of the inner cone (copper) mostly constitutes the jet.

The study can be further extended to design shaped charges for more effective anti-tank ammunitions and oil-well perforation with increased performance.

Symbols and Abbreviations

BMLSC : Bimetallic liner shaped charge
MLSC : Monolithic liner shaped charge
HEAT : High explosive anti-tank
ρ : density

Acknowledgements
We acknowledge Director General, DGAQA New Delhi; Director-College of Engineering,Pune; Director-Armament Research and Development Establishment Pune; Director-High Energy Materials Research Laboratory Pune and Vice Chancellor-Defence Institute of Advanced Technology Pune for their wholehearted support and constant encouragement in the present work.

References

[1] Ernest L Baker, *Modeling and optimization of shaped charge liner collapse ad jet formation*, Technical Report ARAED-TR-92019, US Army Research Development and Engineering Center, New Jersey USA, 1993.

[2] Robert A Brimmer, *Manual for shaped charge design*, Report 1248, US Naval Ordnance Test Station China Lake, CA, USA, 1950.

[3] WP Walters and SK Golaski, *Hemispherical and conical shaped-charge liner collapse and jet formation,* Report BRL-TR-2781, US Army Ballistic Research Laboratory Aberdeen Proving Ground, Maryland, USA, 1987.

[4] David Hasenberg, *Consequences of coaxial jet penetration performance and shaped charge design criteria*, Thesis - NPS-PH-10-0010 , Naval Postgraduate School, CA, USA , 2010.

[5] Saul Skolnick and Albert Goodman, *Energy transfer through a multi-layer liner for shaped charges*, US Patent 4498367, Southwest Energy Group Ltd Albequerque N Mex, 1985.

[6] P.Y. Chanteret and A. Lichtenberger, Bimetallic liners and coherence of shaped charge jets, *Proc. 15th International Symposium on Ballistics*, Jerusalem, Israel, 21-24 May 1995, p.143.

[7] Marinko Ugrcic and Dušan Ugrcic, FEM Techniques in Shaped Charge Simulation, *Scientific Technical Review ,*2009, LVIX ,1, 26-33.

[8] Autodyn User Manual, Century Dynamics Inc., 2001.

[9] DH Lassila, WP Waiters, DJ Nikkel, Jr., RP Kershaw, Analysis Of Soft Recovered Shaped Charge Jet Particles, *Symp. Structures Under Extreme Loading Conditions at 1996 ASMEs Pressure Vessels and Piping Conference*, Montreal, Canada July 21-26, 1996 (Lawrence Livermore National Lab USA, Report No. UCRL-JC-123850 Apr 1996). https://doi.org/10.2172/251380

[10] Steven M Buc, *Liner materials: resources, processes, properties, costs and applications*, SPC-91-282-2, DARPA, Viginia, USA, 1991.

Explosion Shock Waves and High Strain Rate Phenomena
Materials Research Proceedings **13** (2019) 149-153

Materials Research Forum LLC
https://doi.org/10.21741/9781644900338-25

Underwater Explosive Welding of Tin and Aluminium Plates

Satyanarayan[1,a*], Shigeru Tanaka[2,b] and Kazuyuki Hokamoto[3,c]

[1]Department of Mechanical Engineering, Alva's Institute of Engineering and Technology, Moodbidri – 574225, India

[2]Institute of Pulsed Power Science, Kumamoto University, Kumamoto, 8608555, Japan

[3]Institute of Pulsed Power Science, Kumamoto University, Kumamoto, 8608555, Japan

[a]satyan.nitk@gmail.com, [b]tanaka@mech.kumamoto-u.ac.jp, [c]hokamoto@mech.kumamoto-u.ac.jp

Keywords: Underwater Shockwave, Al/Sn, Explosive Welding, interface, Wavy Morphology

Abstract. In the present study, underwater explosive welding of commercial pure Sn and Al plates was attempted. Distance between the explosive and the center of the sample was varied to change the pressure applied to the plates to be welded. Evolution of interfacial microstructures at the welded Sn/Al joints was assessed. An increase in the distance between explosive and the sample exhibited decrease in the formation of wavy morphology at the interface. Cross–sectional interfacial microstructures clearly indicated that, Sn and Al plates can be joined successfully using underwater explosive welding technique.

Introduction

The Explosive welding (EXW) is a solid state process used for the joining (metallurgical) of similar or dissimilar a metal which is regarded as one of the most widely employed materials processing technique [1]. The EXW is generally performed in an open atmosphere. However, it is reported that conventional explosive welding always poses a problem for welding of materials, particularly for thin metal plate (below 1mm thickness) as well as brittle materials such as amorphous ribbon/ceramics and fusing of tungsten (W)/Cu [2,3]. Literature suggested that by using underwater explosive welding a significant decrease in kinetic energy (K.E) loss at the interface of flyer plate and base plate can be achieved [4–6]. In this method, water acts as a pressure transmitting medium. The underwater shock waves prevent the distortion of the welded joint and ensure the integrity of the joints. Hence, underwater explosive welding is regarded as one of the best and novel welding techniques [7, 8]. It reported that, Al/Steel, Al/Cu, Sn/Cu and Cu/Stainless Steel combinations of materials are the most essential in the electrical engineering and among these Al/Cu joints are widely used as electrical connectors in many industries because of their good corrosion resistance and electrical conductivity [9]. Although numerous investigations on explosive welding of various metal combinations were conducted by the researchers [9–12] welding and cladding of Sn and Al using this technique have not been paid attention. Sn based solder alloys are electrically connected with metallic components (most notably the Cu conductors) in the electronic device. However there is no solder alloy in electronic applications which operates with Al in the same way that ordinary solders operate with copper. Because Al does not alloy readily with solders, moreover the Al surface is covered with a thin invisible coating of aluminium oxide. Thin oxide film makes it difficult to join dissimilar materials [13].

Thus, the aim of current study is to make an attempt to fusing of Sn and Al plates using underwater explosive welding method. Further, evolution of interfacial microstructures between welded Sn/Al joint is investigated.

Experimental

The commercial high purity Sn (0.5 mm × 100 mm × 100 mm) and Al (5mm × 100 mm × 100 mm) plates procured from Nilaco corporation, Japan were used in the present study. The procured Sn plate was sectioned into small plates having a dimension of 0.5 mm thick × 50 mm length × 50 mm width and Al plate of 5mm × 50 mm × 50 mm. Underwater explosion welding experiments with an inclined setup were performed to weld Sn and Al plates. A stand–off distance (SOD) between the flyer plate Sn and base plate Al was set to 0.2 mm by placing 0.2 mm thick aluminium plate as spacer between the plates. The inclination angle ($\propto$) between the plates was set to 20° to control the collision angle and the velocity. A stainless steel (SUS 304, 0.1 mm ×50 mm× 50 mm) was used as cover plate above the flyer plate to eliminate cracks between the joints. An inclined layer of SEP explosive (detonation velocity of 7 km/s, density 1300 kg/m³) was bonded to Polymethyl Methacrylate (PMMA) plate and positioned above the flyer plate. The SEP explosive was procured from Kayaku Japan Co, Ltd, Japan. The distance between explosive and the center of the sample (d) were set at 30 and 60 mm. A mild steel anvil was positioned below the sample to ensure the sample flatness and to adjust its height. Entire setup was kept inside PMMA container which contained water. Fig. 1 shows the schematic diagram of underwater shockwave explosion welding technique with weldable conditions.

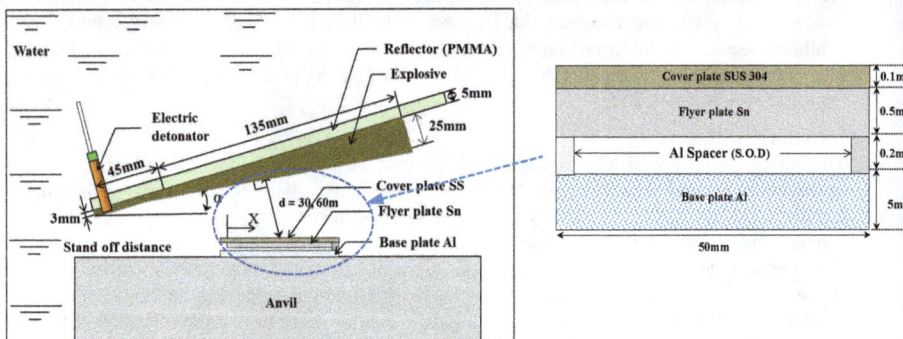

Fig. 1: Schematic outline of explosive welding using underwater shock wave technique.

Welded Sn/Al plates were sectioned along the direction of wave propagation using shear cutting machine (Aizawa, AST–612). Sectioned samples were polished using SiC papers of different grit sizes (400–2000 mesh number) using emery paper disc polishing machine (Velnus, Asahikase make). The final polishing was carried out on a disc polisher (Struers labpol – 1) using silica liquid lubricant. Interfacial region of Sn/Al joint cut at the center parallel to the detonation direction was micro–examined using an optical microscope (Nikon LM 2) and scanning electron microscope (JEOL JSM 6510A).

Results and discussion

Underwater explosive bonded Sn/Al plates at varied distance of the explosive from the center of the sample are shown in Fig. 2. As the detonation initiated, chemical reaction of explosive at high rate generated the shockwaves in the surrounding water. These shockwaves propagated through the water and accelerated the Sn plate (flyer) to impact the Al plate (base). Due to the collision (at higher rate), a strong metallurgical bond between Sn and Al was occurred.

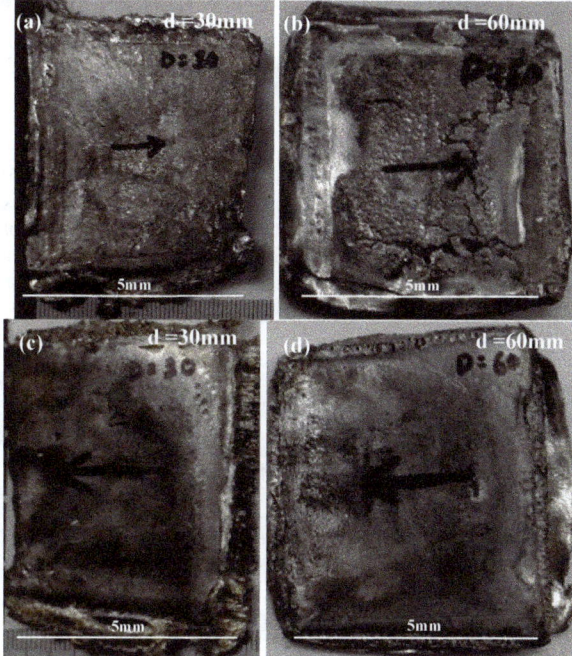

Fig. 2: Explosive welded (underwater) Sn/Alu samples at (a) d=30mm front view (c) d=30mm back view (b) d=60mm front view (d) d=60mm back view

If the distance between explosive and the center of the sample is kept low, collision takes place before the flyer Sn plate could reach the maximum velocity. Further, at higher distance of explosive from the sample, the velocity drops to a lower value at the time of collision [8,14]. The optical microstructures of cross section along welding direction (horizontal positions) for the samples welded at d = 30 mm and 60 mm are shown in Fig. 3 and 4.

Fig 3: Microstructures of cross section along the welding direction (horizontal positions) of welded Sn/Al plates at d = 30 mm.

Fig 4 : *Microstructures of cross section along the welding direction (horizontal positions) of welded Sn/ Al plates at d = 60 mm.*

Interfacial microstructures (at higher magnification) of welded Sn/Al joints at varied water distance are shown in Fig. 5 and 6. Results indicated that pure Sn and Al can be successfully welded/joined using underwater explosive welding method. During explosive welding process, kinetic energy (K.E) of the flyer plate Sn was transformed to potential by colliding with the base plate Al. This resulted in plastic deformation at the interface of Sn and Al plates. Due to higher intensive plastic deformation at d=30mm, a wavy morphology (Fig. 5) was formed at the interface, which swept the surface layers of Sn over Al base plate. However, the kinetic energy of flying plate, energy shockwave and plastic flow was found to be minimum at d = 60 mm due to which small waves in smaller wavelength (Fig. 6) were observed at the Sn/Al interface.

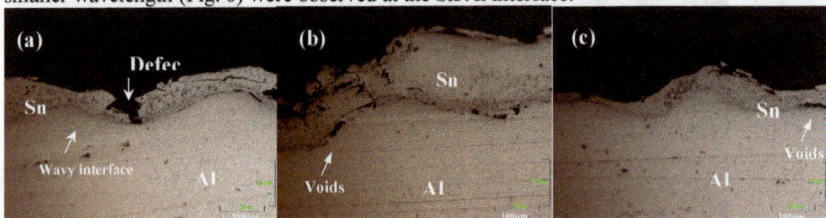

Fig. 5: *Optical microstructures of explosive welded Sn/Al joints (d=30mm) at different locations*

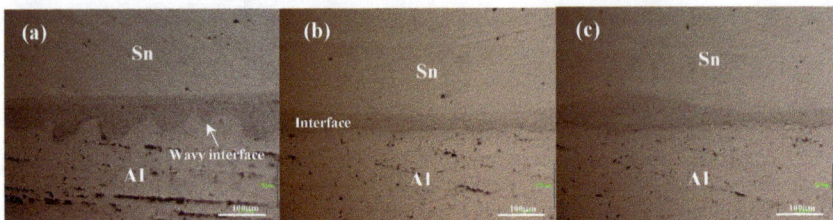

Fig. 6: *Optical microstructures of explosive welded Sn/ Al joints (d=60mm) at different locations*

At the interface of Sn/Al plates welded at d = 30 mm other than large waves, cracks and voids were formed because, the energy shockwave was more, and loss of kinetic energy of flyer plate was high compared to d = 60 mm. Microstructures indicated that the welding was not successful at interface of Sn/Al joints welded at d = 30 mm. However, a good bonding was observed at interface of Sn/Al joints welded at d = 60 mm.

Conclusion
Based on the results and discussion the following conclusions are drawn.
   Sn and Al plates can be successfully bonded using underwater explosive welding method.

Explosion Shock Waves and High Strain Rate Phenomena
Materials Research Proceedings **13** (2019) 149-153

Materials Research Forum LLC
https://doi.org/10.21741/9781644900338-25

The size and morphology of wavy interface decreased with increase in the distance between the explosive and the center of the sample

The interface of samples welded at d = 30 mm associated with the cracks and voids, whereas samples welded at d=60 mm was found to be free defects.

References

[1] B Crossland, J.D.Williams, Explosive welding. Metall. Rev. 15 (1), (1970) 79–100.

[2] H. Iyama, A. Kira, M. Fujita, S. Kubota, K. Hokamoto, S Itoh, An investigation on underwater explosive bonding process. J. Pressure Vessel Technol. Trans. ASME 123 (4) (2001) 486–92. https://doi.org/10.1115/1.1388007

[3] P. Manikandan, J.O. Lee, K. Mizumachi, A. Mori, K. Raghukandan, K. Hokamoto Underwater explosive welding of thin tungsten foils and copper. J. Nucl. Phys.Mater. Sci. Radiat. Appl. 418 (1–3) (201) 281–285. https://doi.org/10.1016/j.jnucmat.2011.07.013

[4] K. Hokamoto, M. Fujita, M., H. Shimokawa, Explosive welding of a thin metallic plate onto a ceramic plate using underwater shock wave. Rev. High Press. Sci. Technol. 7 (1998) 921–923. https://doi.org/10.4131/jshpreview.7.921

[5] A. Mori, K. Hokamoto, M. Fujita, Characteristics of the new explosive welding technique using underwater shock wave-Based on numerical analysis. Mater. Sci. Forum 465 (2004) 307–312. https://doi.org/10.4028/www.scientific.net/msf.465-466.307

[6] A. Mori, K. Hokamoto, M. Fujita, Controlling shock pressure distribution for explosive welding using underwater shockwave. J. JSTP 47 (542) (2006) 195–199 in Japanese. https://doi.org/10.9773/sosei.47.195

[7] D. Mori, K. Ryuta, S. Konishi, Y. Morizono, K. Hokamoto, K Underwater explosive welding of tungsten to reduced-activation ferritic steel F82H. Fusion Eng. Des. 89 (2014) 1086–1090. https://doi.org/10.1016/j.fusengdes.2013.12.038

[8] Satyanarayan, S. Tanaka, A. Mori, K. Hokamoto, K., Welding of Sn and Cu plates using controlled underwater shock wave. J. Mater. Process Technol. 245 (2017) 300–308. https://doi.org/10.1016/j.jmatprotec.2017.02.030

[9] K. Paul, V. Cyril, M. Jukka, S. Raimo, Factors influencing Al-Cu weldproperties by intermetallic compound formation. IJMME 10 (2015) 10.

[10] M. Acarer, B. Gulenc, F. Findik, The influence of some factors on Steel/Steelbonding quality on their characteristics of explosive welding joints. J. Mater.Sci. 39 (21) (2004) 6457 66.

[11] A. Durgutlu, B. Gulenc, F. Findik, Examination of copper/stainless steel joints formed by explosive welding. Mater. Des. 26 (6) (2005) 497–507. https://doi.org/10.1023/b:jmsc.0000044883.33007.20

[12] F. Findik, Recent developments in explosive welding. Mater. Des. 32 (2011.)1081–1093. https://doi.org/10.1016/j.matdes.2010.10.017

[13] Satyanarayan and K N Prabhu., 2011. Wetting behaviour and interfacial microstructure of Sn–Ag–Zn solder alloys on nickel coated aluminium substrates, Materials Science and Technology, 27:7, 1157-1162. https://doi.org/10.1179/026708310x12815992418337

[14] Satyanarayan, A. Mori, M. Nishi, and K. Hokamoto Underwater shock wave weldability window for Sn-Cu plates. Journal of Materials Processing Technology, 267 (2019) pp.152-158. https://doi.org/10.1016/j.jmatprotec.2018.11.044

Explosion Shock Waves and High Strain Rate Phenomena
Materials Research Proceedings 13 (2019) 154-158

Materials Research Forum LLC
https://doi.org/10.21741/9781644900338-26

Development of Weldability Window for Aluminum-Steel Explosive Cladding

G. Murugan*[1,a], S. Saravanan[1,b] and K. Raghukandan[2,c]

[1]*Department of Mechanical Engineering, Annamalai University, Annamalainagar, Tamilnadu, India

[2]Department of Manufacturing Engineering, Annamalai University, Annamalainagar, Tamilnadu, India

[a]geeyemcdm@gmail.com, [b]ssvcdm@gmail.com, [c]raghukandan@gmail.com

Keywords: Explosive Cladding, Aluminum, Stainless Steel, Weldability Window, Microstructure

Abstract. This study addresses the development of biaxial and triaxial weldability window- an analytical estimation-for determining the nature of interface in aluminum 5052-stainless steel 304 dissimilar explosive cladding. The lower and upper boundaries of the biaxial weldability window are formulated using empirical relations proposed by earlier researchers. The process parameters - dynamic bend angle and collision velocity are chosen as ordinates and abscissa respectively. In addition, a triaxial weldability window, comprising of three process parameters viz., flyer plate velocity, collision velocity and dynamic bend angle is also developed. Explosive cladding experiments were conducted by varying the process parameters and the interface microstructure is correlated with the developed weldability windows.

1. Introduction

Development of bimetallic components drew the attention IN aerospace, ship building and automotive sectors, owing to their lightweight and good mechanical, thermal and corrosion resistance properties [1]. Aluminum clad steel is one such bimetallic component, used, as transition joint, in cryogenic pressure vessels and in power station cooling system. Numerous techniques are currently employed for cladding aluminum with steel such as hot rolling, hot pressing and diffusion bonding. However due to wide difference in density, melting point and coefficient of thermal expansion, bonding of aluminum to steel by conventional welding methods, is still a challenging one [2]. In this context, explosive cladding shows good potential to manufacture larger bimetallic clads and composite laminates without complications.

In explosive cladding, the quality of the clad depends on the judicial selection of process parameters viz., surface preparation, standoff distance, loading ratio, thickness of flyer plate and the properties of the chemical explosive [3]. Various researchers employed numerical simulation and weldability window-an analytical estimation for obtaining the optimum conditions-for an acceptable clad exhibiting microstructure free from defects and good strength [4-6]. In this study, an attempt is made to develop a triaxial weldability window in addition to the conventional biaxial one,for attaining optimum process parameters for a sucessful Al 5052-SS 304 explosive clad. The results show that the experimental conditions prevailing inside the boundaries of the window results in a successful clad with a wavy topography.

2. Experimental

Parallel and inclined explosive cladding configuration, shown in Fig.1 [7], with aluminum 5052 alloy (size: 90 mm × 50 mm × 2 mm) and SS 304 (size:90 mm × 50 mm × 6 mm) as flyer and base plates respectively was attempted.

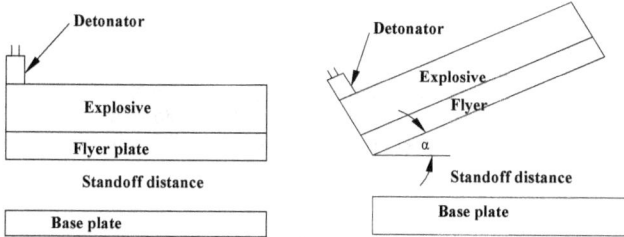

Fig.1 (a) Parallel configuration   (b) Inclined configuration

The mating surfaces of the plates were mechanically polished and thoroughly cleaned prior to experiments. The explosive (detonation velocity of 4000 m/s), was positioned above the flyer plate and the preset angle between the flyer and base plate has varied from 0 to 5 degree. The loading ratio (mass of the explosive/mass of flyer plate) was varied from 0.8 to 1.0 and the detailed experimental conditions are tabulated in Table 1. Subsequent to cladding, samples were cut parallel to the detonation direction for microstructural examination. Microstructural features of the clad were observed under VERSAMET optical microscope and the results are reported.

Table 1 Experimental conditions

S. No	Loading ratio, R	Standoff distance, S [mm]	Preset angle, A [degree]	Dynamic bend angle, β [degree]	Collision velocity, V_c, [m/s]	Flyer plate velocity, V_p [m/s]
1	0.8	5	0	10.03	4000	699.4
2	0.9	5	0	10.89	4000	759.7
3	1.0	5	0	11.7	4000	816
4	0.8	5	3	13.03	3071.7	699.4
5	0.9	5	5	15.89	2728.8	759.7

3. Weldability window

The determination of precise boundaries in the construction of weldability window involves various assumptions and constants which are influential during the formulation.. The experimental conditions prevailing within the upper and lower boundaries of the window, results in a successful clad [8].. In explosive cladding, as the number of process parameters are more, weldability window is drawn between any two chosen influential process parameters. In this study, the biaxial weldability window is generated with welding velocity and dynamic bend angle as ordinate and abscissa respectively. In parallel configured explosive cladding, the welding velocity, V_c, is equal to detonation velocity (V_d) of the explosive [8]. The second chosen parameter dynamic bend angle, β, is analytically determined by

$$\beta = 2\sin^{-1}\frac{V_p}{V_d} \qquad (1)$$

Where V_p is the flyer plate velocity determined by $V_p = 2V_d Sin\frac{\beta}{2}$ (2)

The lower boundary of the weldability window is determined by [4]

Explosion Shock Waves and High Strain Rate Phenomena   Materials Research Forum LLC
Materials Research Proceedings 13 (2019) 154-158   https://doi.org/10.21741/9781644900338-26

$$\beta = K \sqrt{\frac{H_V}{\rho V_C^2}} \qquad\qquad (3)$$

Where, K is equal to 1.14, 'H_v' is the Vickers hardness and 'ρ' is the density of the flyer plate. The experimental conditions, superimposed on the biaxial weldability window, are falling closer to the lower boundary, as shown in Fig.2. The upper boundary of the weldability window is estimated by

$$\sin\frac{\beta}{2} = \frac{K_3}{\left(t^{0.25}.V_C^{1.25}\right)} \qquad\qquad (4)$$

Where $k_3 = C_f/2$, $C_f = \sqrt{K/\rho}$, K= E/3(1-2γ), Where C_f is compressive wave velocity, t is the thickness of flyer plate, V_c is the collision point velocity, k is the bulk modulus and E is the young's modulus.

4. Results and discussion
4.1 Biaxial weldability window
The biaxial weldability window for aluminium-steel, comprising of upper, lower boundaries and the experimental conditions (Table 1), is shown in Fig. 2. Researchers opined that the experimental conditions inside the upper and lower boundaries results in successful clad. However, the significance of regions near to lower boundary, and that too in close proximity to left corner in achieving a defect free clad is insisted by many earlier researchers [4,6,9]. Experimental conditions closer to the lower boundary indicate lower dynamic bend angle, collision velocity and plate velocity and which results in a defect free dissimilar clad (detailed in the section 4.3).

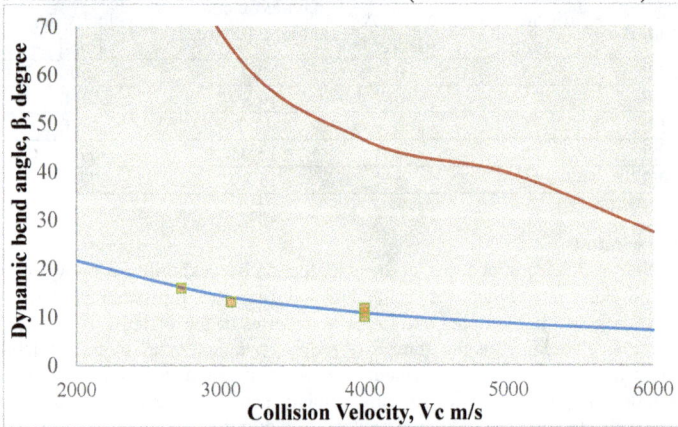

Fig.2 Biaxial weldability window (Al-SS 304)

4.2 Triaxial Weldability window
The lower boundary of the weldability window is generated in a three dimensional view, considering three parameters viz., bend angle, collision velocity and flyer plate velocity (Fig.3). A three dimensional lower boundary in a weldability window provides better understanding of the collision condition as the additional third parameter is considered as well.

Materials Research Forum LLC
https://doi.org/10.21741/9781644900338-26

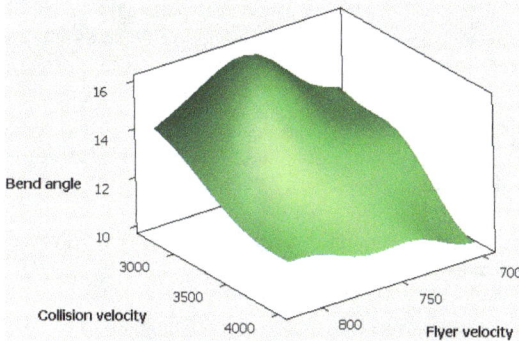

Fig.3 Triaxial weldability window (Al-Steel)

4.3 Microstructure

The interface highlights the difference in microstructure, from straight and sinusoidal topographies, and indicates the effect of explosive mass on the quantum of deformation work performed. Transformation of straight interface to an undulating interface, for an increase in loading ratio, is consistent with earlier researchers [10-12]. None the less, interfacial melting is witnessed at few regions of the crest of the wave for all attempted conditions, due to enhanced temperature at specific locations. Further, grains across the periphery are finer and oriented towards the detonation direction.

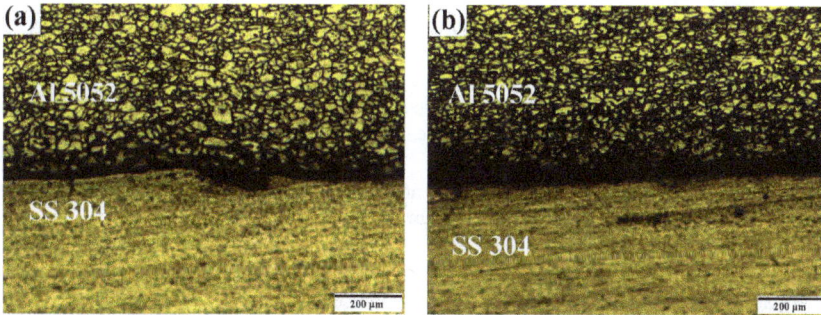

Microstructure of the explosive clad (Al 5052-SS 304)

The interface microstructure of Al-SS 316 explosive clad, for a loading ratio, R of 0.8, show a straight interface with a continuous strip of molten diffusion layer (Fig.4.b). The molten layer, probable weaker locations in the clads, are formed due to the dissipation of the available kinetic energy at the interface. The microstructure of the aluminum 5052-stainless steel explosive clad at a loading ratio, R=1.0 (Fig.4.a) reveals the characteristic wavy interface with few intermetallic compounds on the vortices of the participant metals. The experimental conditions prevailing inside the weldability window results in a wavy interface.

Conclusion

1. Weldability window is a effective tool for selecting explosive cladding parameters.
2. The experimental conditions falling within the window produce a wavy interface.

Explosion Shock Waves and High Strain Rate Phenomena
Materials Research Proceedings 13 (2019) 154-158

Materials Research Forum LLC
https://doi.org/10.21741/9781644900338-26

3. Points closer to the lower limits of welding window are preferable.

4. Triaxial weldability window provides the influence of three process parameters, and hence advantageous.

References

1. Y.Aizawa, J. Nishiwaki,, Y.Harada, S.Muraishi,. and S.Kumai, 2016. Experimental and numerical analysis of the formation behavior of intermediate layers at explosive welded Al/Fe joint interfaces. Journal of Manufacturing Processes, 24, pp.100-106.
https://doi.org/10.1016/j.jmapro.2016.08.002

2. S.Saravanan. and K.Raghukandan, 2011. Energy dissipation in explosive welding of dissimilar metals. Materials Science Forum, 673, pp. 125-129.
https://doi.org/10.4028/www.scientific.net/msf.673.125

3. F.Findik, 2011. Recent developments in explosive welding. Materials & Design, 32(3), pp.1081-1093. https://doi.org/10.1016/j.matdes.2010.10.017

4. Satyanarayan, A.Mori, M.Nishi. and K.Hokamoto, 2019. Underwater shock wave weldability window for Sn-Cu plates. Journal of Materials Processing Technology, 267, pp.152-158.
https://doi.org/10.1016/j.jmatprotec.2018.11.044

5. X.Wang, Y.Zheng, H.Liu, Z.Shen, Y.Hu, W.Li, Y.Gao. and C.Guo., 2012. Numerical study of the mechanism of explosive/impact welding using smoothed particle hydrodynamics method. Materials & Design, 35, pp.210-219. https://doi.org/10.1016/j.matdes.2011.09.047

6. S.Somasundaram, R.Krishnamurthy. and H.Kazuyuki 2017. Effect of process parameters on microstructural and mechanical properties of Ti− SS 304L explosive cladding. Journal of Central South University, 24(6), pp.1245-1251 https://doi.org/10.1007/s11771-017-3528-3

7. S.Saravanan. and K.Raghukandan, 2012. Thermal kinetics in explosive cladding of dissimilar metals. Science and Technology of Welding and Joining, 17(2), pp.99-103.
https://doi.org/10.1179/1362171811y.0000000080

8. S.Saravanan and K.Raghukandan, 2013. Influence of interlayer in explosive cladding of dissimilar metals. Materials and Manufacturing Processes, 28(5), pp.589-594.
https://doi.org/10.1080/10426914.2012.736665

9. S.A.Mousavi. and P.F.Sartangi, , 2009. Experimental investigation of explosive welding of cp-titanium/AISI 304 stainless steel. Materials & Design, 30(3), pp.459-468.
https://doi.org/10.1016/j.matdes.2008.06.016

10. P.Tamilchelvan, K.Raghukandan. and S.Saravanan, 2014. Kinetic energy dissipation in Ti-SS explosive cladding with multi loading ratios. Iranian Journal of Science and Technology. Transactions of Mechanical Engineering, 38(M1), p.91-96.

11. S.Saravanan, K.Raghukandan. and K.Hokamoto, 2016. Improved microstructure and mechanical properties of dissimilar explosive cladding by means of interlayer technique. Archives of Civil and Mechanical Engineering, 16(4), pp.563-568.
https://doi.org/10.1016/j.acme.2016.03.009

12. I.A.Bataev, D.V.Lazurenko, S.Tanaka, K.Hokamoto, A.A.Bataev, Y.Guo. and A.M.Jorge Jr 2017. High cooling rates and metastable phases at the interfaces of explosively welded materials. Acta Materialia, 135, pp.277-289. https://doi.org/10.1016/j.actamat.2017.06.038

Explosion Shock Waves and High Strain Rate Phenomena
Materials Research Proceedings **13** (2019) 159-162

Materials Research Forum LLC
https://doi.org/10.21741/9781644900338-27

Effect of Silicon Carbide Particles in Explosive Cladded Aluminum Hybrid Composites

S. Saravanan[1,a*], K. Raghukandan[2,b] and G. Murugan[3,c]

[1*]Department of Mechanical Engineering, Annamalai University, Annamalainagar, India

[2]Department of Manufacturing Engineering, Annamalai University, Annamalainagar, India

[3]Department of Mechanical Engineering, Annamalai University, Annamalainagar, India

[a]ssvcdm@gmail.com, [b]raghukandan@gmail.com, [c]geeyemcdm@gmail.com

Keywords: Explosive Cladding, Aluminum, Silicon Carbide, Microstructure, Hardness

Abstract. In this study, explosive cladding of aluminum 5052-aluminum 1100 plates with silicon carbide particles spread between them is attempted. The percentage of silicon carbide particles is varied from 6% to 12% by wt., keeping other parameters viz., standoff distance, loading ratio, flyer and base plate thickness and preset angle as constant. The influence of silicon carbide (SiC_p) on the interface microstructure and strength are discussed and correlated with the conventional explosive clad. The interface microstructure reveals a smooth interface free from defects and an increase in silicon carbide particles enhances the strength of the dissimilar aluminum explosive clad.

Introduction

Hybrid aluminium composites, having one or more reinforcements, are extensively employed in aerospace, marine, automotive and defence applications. Hybrid composites exhibit superior mechanical properties viz., higher hardness, fracture toughness, fatigue resistance and creep resistance [1]. In addition, these composites show higher corrosion and wear resistance and thereby, are excellent replacement for conventional metals and alloys [2]. The choice of processing condition, fabrication process and the choice of reinforcement significantly improve the mechanical and metallurgical properties of hybrid composites.

Various researchers prepared and analysed the metallurgical and mechanical properties of hybrid composites through powder metallurgy techniques. Few of the salient contributions are summarized here. Rana *et al.* prepared silicon carbide reinforced aluminium alloy composites by melt-stir casting assisted with ultrasonic vibrations [3]. They concluded that the ultrasonic vibration improves the grain refinement and promote uniform distribution of the reinforcements. Afkham *et al.* employed aluminium nano particles as reinforcement in the preparation of aluminium matrix composites and reported better mechanical properties [4]. In this context, Ma *et al.* recommended the usage of smaller grain sized reinforcements for attaining high wear and mechanical properties [5]. In another study, Dasari *et al.* examined the mechanical properties of liquid infiltrated aluminium alloys reinforced with graphene oxide and reduced graphene oxides [6]. Shirvanimoghaddam *et al.* studied the improvement in hardness and tensile strength of boron carbide, titanium diboride and zirconium silicate reinforced aluminium alloy composites [7]. Though significant research on hybrid aluminium composites was performed in powder metallurgy technique, studies on the preparation of silicon carbide reinforced dissimilar aluminium composites by explosive cladding is scarce. In this study, Al 5052 and Al 1100 aluminium plates are explosively cladded with varied percentage of silicon carbide particles spread on the base plate. The influence of silicon carbide on microstructure and mechanical strength of the dissimilar aluminium clad is reported.

Experimental

A parallel explosive cladding configuration with silicon carbide particles placed at the base plate was attempted (Fig. 1). Silicon carbide particles of varied proportion (6-12 %, by wt. of flyer plate) are dispersed in the mid region of the base plate (Al 1100), having dimension 80 mm X 60 mm X 5 mm. Al 5052 plate (80 mm X 60 mm X 2 mm) is positioned 5 mm above the base plate (standoff distance).The chemical composition of parent metals are given in Table 1. Commercial chemical explosive (detonation velocity-4000 m/s, density-1.2 g/cm^3) was packed above the flyer plate at a constant loading ratio, R of 0.8, and the detonator was positioned on one corner of the explosive pack. The experimental conditions were fixed based on the trial experiments.

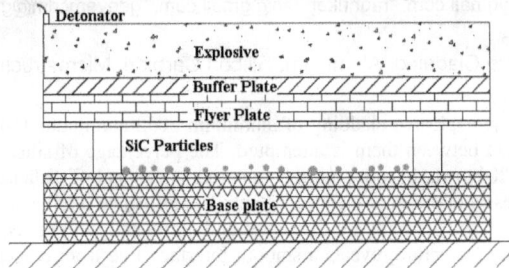

Fig.1 Explosive cladding with silicon carbide particles

Table 1 Chemical composition of participant metals

Material	Composition (wt. %)								
	Cu	Mn	Si	Mg	Zn	Fe	Cr	Ti	Al
Al 1100	0.0292	0.0177	0.101	0.0169	0.0158	0.479	-	-	Bal
Al 5052	0.1	0.4	0.4	4.2	0.25	0.4	0.15	0.15	Bal

Post cladding, the specimens for metallographic observations were sectioned parallel to the detonation direction, following standard metallurgical procedures viz., grinding, polishing and etching (Kellers reagent-5 ml HF, 10 ml H_2SO_4, 85 ml H_2O for 20 s). The post cladding metallographic analysis was performed in a VERSAMAT–3 optical microscope equipped with Clemex image analyzing system. Vickers micro-hardness measurement across the explosive clads were conducted on a ZWICK micro-hardness tester applying 100 g load (ASTM E 384 standard) and the results are presented.

Results and discussion

Microstructure. The interface microstructure of the SiCp dispersed hybrid aluminum explosive clad are shown in Fig. 2(a-d). The interface exhibit a smooth interface free from jet trapping or formation of any reaction compounds. The interfaces reveal the dispersion of silicon carbide particles (SiCp), visible as a continuous black patch. The thickness of black region increases with the percentage of silicon carbide employed at the middle of the dissimilar aluminum clad interface. The silicon carbide particles, with reduced size, are distributed uniformly across the interface. The grains closer to the interface are smaller and oriented towards the direction of detonation.

Explosion Shock Waves and High Strain Rate Phenomena
Materials Research Proceedings **13** (2019) 159-162

Materials Research Forum LLC
https://doi.org/10.21741/9781644900338-27

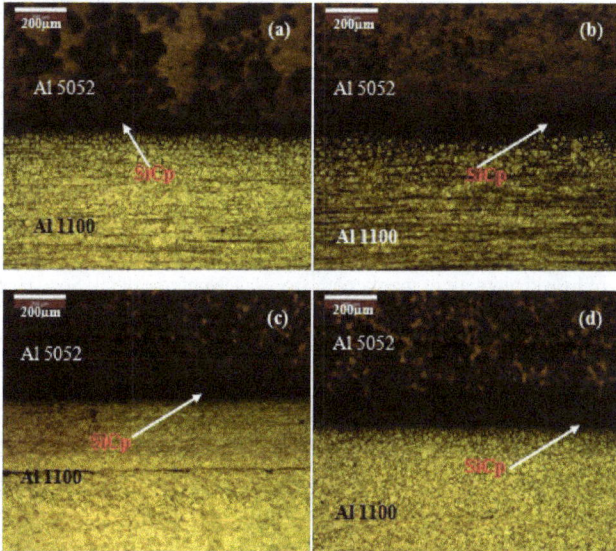

Fig. 2 Microstructure of the Al5052-Al 1100 explosive clad
(a) 6 % SiCp (b) 8 % SiCp (c) 10 % SiCp (d) 12 % SiCp

The interface microstructure of dissimilar aluminium (Al 5052-Al 1100) explosive clad with 6% of SiCp is shown in Fig. 2.a. It is observed that the silicon carbide particles are uniformly distributed across the interface. The interface exhibit a straight interface, free from trapping of escaping jet or formation of continuous molten interlayer. The thickness of silicon carbide particles at the interface is measured as 1.39 mm. However, it increases to 1.48 mm, when the quantity of silicon carbide is enhanced to 8 % (Fig. 2.b). Formations of cluster or agglomerations of SiCp particles are not observed at the interface (Fig.2.b). The SiCp layer's thickness at the dissimilar aluminium clad interface further increases to 1.54 mm, when the concentration reaches 10 % (Fig.2.c). On the contrary, for a 12 % SiCp concentration, the micrographs (Fig. 2.d) reveal agglomeration and particle clusters of Al_2SiC, identified by EDS analysis. Formation of agglomeration or clusters is consistent with the reports of Dhas et al [8]. The concentration of silicon carbide particles at the interface holds a significant effect on the micro-hardness of the interface as well (detailed in the next section).

Hardness. The average of three Vickers micro-hardness measurements at equal intervals from the cross sections of silicon carbide reinforced dissimilar aluminum explosive clad is shown in Fig.3. Post clad hardness of both aluminum grades are higher than the pre clad conditions (Al 1100- 47 Hv, Al 5052-74Hv) following high pressure and the cold deformation experienced during the high impact collision. This is consistent with the earlier reports of Saravanan *et al* [9-11]. There is no significant variation in the hardness at the hardness of silicon carbide particles.

Conclusions
The present study reports the influence of varied percentage of silicon carbide on the microstructure and micro-hardness of Al 5052-SiCp-Al 1100 explosive clads. The following salient conclusions are drawn from this experimental study:

1. The introduction of silicon carbide as reinforcement in aluminium composites significantly improves the strength characteristics.
2. The Al 5052-SiCp-Al 1100 explosive clad with 10 % silicon carbide exhibit better strength.
3. As the concentration of silicon carbide particles enhances its thickness on the interface increases as well.
4. The enhancement in the mechanical properties of silicon carbide reinforced dissimilar aluminium explosive clads promises to be a potential candidate for marine and offshore structural applications

References

[1] T. Rajmohan, T.K. Palanikumar, S. Ranganathan, Evaluation of mechanical and wear properties of hybrid aluminium matrix composites. T.Nonferr. Metal. Soc. 23(9) (2013) 2509-2517. https://doi.org/10.1016/s1003-6326(13)62762-4

[2] S. Saravanan, K.Raghukandan, P. Kumar, Effect of wire mesh interlayer in explosive cladding of dissimilar grade aluminum plates, J.Cent. South Univ, 26 (3) (2019) 604-611. https://doi.org/10.1007/s11771-019-4031-9

[3] R.S. Rana, R. Purohit, V.K. Soni, S. Das, Characterization of mechanical properties and microstructure of aluminium alloy-SiC composites. Mater. Today 2(4-5), (2015)1149-1156. https://doi.org/10.1016/j.matpr.2015.07.026

[4] Y. Afkham, R.A.Khosroshahi, S.Rahimpour, C.Aavani, D.Brabazon, R.T.Mousavian, Enhanced mechanical properties of in situ aluminium matrix composites reinforced by alumina nanoparticles. Arch. Civil Mech. Engg, 18(1), (2018) 215-226. https://doi.org/10.1016/j.acme.2017.06.011

[5] S.Ma, E.Xu, Z.Zhu, Q. Liu, S.Yu, J.Liu, H.Zhong, Y. Jiang, Mechanical and wear performances of aluminum/sintered-carbon composites produced by pressure infiltration for pantograph sliders. Powder Technol. 326 (2018) 54-61. https://doi.org/10.1016/j.powtec.2017.12.027

[6] B.L.Dasari, M. Morshed, J.M.Nouri, D. Brabazon, S.Naher, Mechanical properties of graphene oxide reinforced aluminium matrix composites. Compos Part B: Eng, 145 (2018)136-144. https://doi.org/10.1016/j.compositesb.2018.03.022

[7] K. Shirvanimoghaddam, H.Khayyam, H.Abdizadeh, M.K.Akbari, A.H. Pakseresht, F. Abdi, A. Abbasi, M. Naebe, Effect of B_4C, TiB_2 and $ZrSiO_4$ ceramic particles on mechanical properties of aluminium matrix composites: Experimental investigation and predictive modelling. Ceram. Int., 42(5) 2016, 6206-6220. https://doi.org/10.1016/j.ceramint.2015.12.181

[8] D.E.J. Dhas, C. Velmurugan, K.L.D.Wins, K.P.BoopathiRaja, Effect of tungsten carbide, silicon carbide and graphite particulates on the mechanical and microstructural characteristics of AA 5052 hybrid composites. Ceram. Int. 45(1), (2019) 614-621. https://doi.org/10.1016/j.ceramint.2018.09.216

[9] S. Saravanan, K.Raghukandan, K.Hokamoto, Improved microstructure and mechanical properties of dissimilar explosive cladding by means of interlayer technique. Arch.Civil Mech. Engg. 16(4) (2016) 563-568. https://doi.org/10.1016/j.acme.2016.03.009

[10] S. Saravanan, K.Raghukandan, Diffusion kinetics in explosive cladding of dissimilar alloys as described through the Miedema model. Arch. Metall. Mater. 59(4) (2014) 1615-1618. https://doi.org/10.2478/amm-2014-0274

[11] S. Somasundaram, R.Krishnamurthy, H.Kazuyuki, 2017. Effect of process parameters on microstructural and mechanical properties of Ti– SS 304L explosive cladding. J. Cent. South Univ. 24(6) (2017) 1245-1251. https://doi.org/10.1007/s11771-017-3528-3

Explosion Shock Waves and High Strain Rate Phenomena
Materials Research Proceedings 13 (2019) 163-167

Materials Research Forum LLC
https://doi.org/10.21741/9781644900338-28

Microstructural and Mechanical Properties of Al 5052-SS 316 Explosive Clads with Different Interlayer

E. Elango*,1,a, S. Saravanan1,b, K. Raghukandan2,c

1Department of Mechanical Engineering, Annamalai University, Tamilnadu, India

2Department of Manufacturing Engineering, Annamalai University, Tamilnadu, India

aeelango69@gmail.com, bssvcdm@gmail.com, craghukandan@gmail.com

Key words: Explosive Cladding, Aluminium, Steel, Microstructure, Strength

Abstract. This study focuses on the effect different interlayer viz., copper, aluminium and stainless steel interlayer on the explosive cladding of aluminum alloy (Al 5052) - stainless steel (SS 316) plates subjected to varied process parameters viz., standoff distance, loading ratio (mass of explosive/mass of flyer plate) and inclination angle. The interface transforms from straight to wavy, while increasing the standoff distance and loading ratio. Moreover, increase in loading ratio enhances the wave length and the amplitude of interfacial wave. Mechanical testing viz., Vickers micro-hardness, Ram tensile and side shear test were conducted on Al 5052-SS 316 explosive clads and the results are reported. The maximum hardness is obtained for Al-SS 304-SS 316 explosive clads, while the tensile and shear strength of aluminum-SS 316 explosive clads with copper interlayer exhibit an acceptable joint strength.

1. Introduction

Aluminium - steel bimetals are employed in ship building and as high speed transition joints owing to light weight and good corrosion resistance [1]. Aluminium - steel clad plates replace solid aluminium or steel in structural, thermal expansion management and corrosion resistant applications. The reason for using aluminium - steel composite part instead of single metal is to lower cost with better corrosion resistance and improved strength. Welding of aluminium-steel plates by conventional methods is not viable due to the formation of undesirable intermetallic compounds, which weakens the clad strength and results in a poor metallurgical bond. Whereas, explosive cladding offer a feasible alternative to clad aluminium-steel plates devoid of intermetallic compounds at minimum cost. The quality of explosive clad is dictated by the proper selection of process parameter viz., standoff distance, loading ratio and inclination angle [2-4].

Acarer et al. investigated the effects of process parameters (explosive rate, anvil, stand -off distance) on microhardness and shear strength of the dissimilar explosive clad [5]. In another study, Raghukandan [6] adopted Response Surface Methodology to evaluate the effect of process parameters viz., flyer thickness, explosive loading ratio (R), angle of inclination and standoff distance on the tensile and shear properties of Cu-low carbon steel explosive clads. The effect of heat treatment on the aluminium-steel clad strength was reported by Mousavi et al [7]. Recently, Saravanan et al. employed different layer in Al-Cu explosive cladding and reported the significance of kinetic energy utilization on the nature of interface and strength [8]. Similarly, Tamilchelvan et al. cladded titanium-steel at varied loading ratios and standoff distances and who reported the significance of kinetic energy dissipation [9]. Though numerous attempts were made by earlier researchers to explosively clad dissimilar metals, the studies on the effect of different interlayer on Al-steel explosive cladding is limited, and attempted herein. In addition, the mechanical strength of Al-SS 316 explosive clads with different interlayer is determined experimentally, as per the relevant standards, and the results are reported.

Explosion Shock Waves and High Strain Rate Phenomena   Materials Research Forum LLC
Materials Research Proceedings **13** (2019) 163-167   https://doi.org/10.21741/9781644900338-28

2. Experimental Procedure

Inclined explosive cladding configuration reported elsewhere [10] was attempted with aluminum 5052 (50 mm × 100 mm × 2 mm and SS 316 (50 mm × 100 mm × 6 mm) as flyer and base plate respectively. The interlayers viz. Copper (chemical composition in wt%: Mn-0.0002, Si-0.0004, Mg-0.0001, Zn-0.00042, Fe-0.032, Al-0.001, Cu-Bal.), aluminium (chemical composition in wt%: Cu-0.0292, Mn-0.017, Si-0.101, Mg-0.0169, Zn-0.0158, Fe-0.479, Al-Bal) and SS 304 (chemical composition in wt%: Cr-18, Ni-8, Cu-0.05, C-.08, Si-0.34, Mo-0.05, Mn-2, P-0.04,S-0.03, Fe-Bal.) are positioned as interlayer between flyer and base plates. The flyer-interlayer and interlayer-base plates are separated by 10 mm, which allow the flyer plate to reach its terminal velocity. A constant loading ratio and inclination angle (R-1.0 &A-10°) are maintained and the detailed experimental conditions are given in Table 1. The chemical explosive (detonation velocity 4,000 m/s, density 1.2 g/cm^3) was packed above the flyer plate, and the detonator was positioned on one corner. The mating surfaces were mechanically polished and thoroughly cleaned by acetone, prior to experiments.

Post cladding, the clads were sectioned parallel to the detonation direction for examining the nature of interface, and the samples were prepared through standard metallographic practice. Vickers micro-hardness was measured based on ASTM E 384 standard [11] on a ZWICK micro-hardness tester with a load of 4.9 N and a dwelling period of 0.5 mm/min. The averages of three hardness values are values are presented. Ram tensile test specimens for each experimental conditions were prepared in the direction of detonation (MIL-J-24445A standard) and shear test specimens were fabricated as per ASTM B898-99 standard. Both the tests were performed in a servo controlled universal testing machine (UNITEK-94100) by applying uni-axial compressive force on the explosive clads and the results are reported.

		Table 1 – Experimental conditions			
No	Inter layer	Standoff distance, SD, mm	Loading ratio, R	Inclination angle, A degree	Kinetic energy loss,ΔKE, MJm^{-2}
1	Cu	10	1.0	10	0.76
2	Al	10	1.0	10	0.82
3	SS 304	10	1.0	10	0.82

3. Results and Discussion

3.1 Microstructural Characterization

The interface microstructure of Al 5052-SS 316 explosive clad with copper, aluminium and stainless steel (SS 304) interlayers (Fig. 1a-c) show wavy morphologies as reported by earlier researchers [12, 13]. Formation of straight interface is observed on the similar metal sides, whereas, they transform into a wavy interface on the dissimilar side. The undulating interfaces, a noticeable characteristic of explosive cladding process, provide a better interlocking mechanism as the interfacial morphologies are designed and regulated by the system parameters viz., collision angle, collision velocity, preset angle, nature of explosive, standoff distance and properties of participant metals.

The Al 5052-Cu-SS 316 (Fig. 1a) microstructure display a wavy interface devoid of defects, viz., cracks, trapped jet and molten layered zone. When copper is introduced as interlayer between Al-5052 and SS 316 clad, the interfacial waves (amplitude-27µm) are more pronounced on the first interface (flyer-interlayer), whereas the amplitude of interfacial waves (20 μm) declines on the second interface (interlayer-base).

The Al5052-SS316 clad with aluminium interlayer display a straight interface with devoid of crakes, trapped jet and molten layer. When aluminium interlayer is introduced, the

Fig. 1 (a-c) Microstructure of the Al-Steel explosive clad
(a) Cu interlayer (b) Al interlayer (c) SS 304 interlayer

available kinetic energy (0.82MJm^{-2}) increases, thereby metal flow around the collision point becomes unstable and oscillates, creating a wavy interface with a higher amplitude (Fig. 1b: 32 μm) consistent with the report of Somasundaram et al[14]. The kinetic energy spent during collision in explosive cladding with interlayer is given by an empirical relation [8]

$$\Delta KE = \frac{m_f \, m_b \, V_{p2}}{2(m_{f+m_i})} + \frac{M \, m_b \, V_{p1}^2}{2(M + m_b)} \tag{1}$$

Where 'm_f' is the mass of flyer plate per unit area, 'm_i' is the mass of interlayer per unit area, 'm_b' is the mass of base plate per unit area,' V_{p1}' is the flyer plate velocity and M is the combined mass of flyer plate and interlayer. The flyer plate velocity (V_{p1}) after the first impact is calculated by

$$V_{p1} = 2V_d \sin (\beta/2) \tag{2}$$

where 'β' is the dynamic bend angle, calculated by

$$\beta = \left(\sqrt{\frac{k+1}{k-1}} - 1 \right) \cdot \frac{\pi}{2} \cdot \frac{R}{(r+2.71+0.184t_e/SD)} \tag{3}$$

Where R is the loading ratio, 'te' is the thickness of explosive and 'SD' is the standoff distance. K is a constant varies from 1.96 to 2.8 depends on the thickness of the explosive.

The Al 5052-SS304-SS316 (Fig.1c) shows a straight interface devoid of intermetallic compounds and amplitude of explosive clad reduces to resemble a straight interface, when a lower thermal diffusivity stainless steel 304 (4.03 x 10^{-6} m^2/s) is used as interlayer. Though wavy interfaces are preferred, straight interface provides better strength as reported by Kahraman et al. [15]. Hence, it is consistent in this study. The nature of interface viz., wavy, straight or formation of intremetallic compounds is also established by the thermal diffusivity (α) of the interlayer and defined by

$$\alpha = \frac{k}{\rho c} \tag{4}$$

Where, k, ρ and c denote thermal conductivity, density and specific heat capacity of metals respectively [4].

3.2 Mechanical Strength
3.2.1 Microhardness test

The Vicker microhardness of the explosively cladded Al5052-SS316 plates with different interlayer are measured at uniform interval. The Vickers hardness closer to the interface of interlayered clad is twice higher and (20%) more than the base metals. This is because of interface hardening owing to sudden deformation. The increase of hardness in the base plate closer to the interface is expressed and frequently discussed by earlier researchers [5,9, 10&16]. The hardness profile for Al5052-

SS316 explosive clads with different interlayer emphasizes the significance of higher density interlayer. The highest hardness is achieved in stainless steel (SS 304) interlayer following the higher collision velocity. This causes strong plastic deformation due to higher kinetic energy utilization. This enhancement in hardness is not significant at region away from interface following the reduction in plastic deformation.

3.2.2 Ram tensile test

Ram tensile strength of clads are higher than the weaker parent metal (Al5052-180MPa). The lowest tensile strength is obtained (263MPa) when aluminium is employed as interlayer however it is 20% higher than weaker parent metal. The introduction of higher density metals enhances the kinetic energy utilization and influences the mechanical strength of explosive clads. The highest tensile strength value is obtained for experimental condition involving stainless steel (SS 304) interlayer (278MPa) shown in Table 2. Mastanaiah et al. opined that ram tensile strength of explosive clads are invariably higher than the weaker of the participating metals, which is consistent with this study [17].

Table 2 – Tensile strength of explosive clads

No	Standoff Distance (mm)	Loading Ratio (R)	Inclination Angle (A^0)	Interlayer	Tensile strength (MPa)	Shear Strength (MPa)
1	10	1.0	10	Al	263	180
2	10	1.0	10	Cu	273	184
3	10	1.0	10	SS 304	278	186

3.2.3 ASTM side shear strength

The compressive force is applied on the explosive clad by measuring load with respect to displacement, until the sample fails on universal testing machine at 0.5 mm/min and the results are shown in Table 2. The shear fracture took place in the weaker metal (Al 5052) indicating the interface has higher strength than the weaker metal. This is in agreement with the results on Mousavi et al. who joined Ti-steel. [18]. The maximum shear strength is obtained at Al5052-SS304-SS316 explosive clad due to higher kinetic energy utilization at the interface. The shear strength of the clads are 60% - 67% higher than the aluminum and prevailing between the shear strengths of parent metals, which is consistent with the reports of Rao et al [19].

Conclusions

1. Introduction of interlayer significantly increases the kinetic energy utilization, and thereby, the formation of intermetallic compounds at the interface is inhibited.
2. Microhardness closer to the interface is higher owing to the sudden deformation experienced.
3. Al5052-SS316 explosive clad with stainless steel interlayer exhibit better mechanical strength.
4. Ram tensile strength and shear strengths is higher than that of Al 5052 indicating the bond is stronger than the parent metals

References

[1] Tricarico, R. Spina, D. Sorgente, M. Brandizzi, Effect of heat treatments on mechanical properties of Fe/Al explosion-welded structural transition joints, Mater. Des. 30 (2009) 2693-2700. https://doi.org/10.1016/j.matdes.2008.10.010
[2] S. Mroz, G, Stradomski, H. Dyja, Using the explosive cladding method for production of Mg-Al-bimetallic bars, Arch. Civ. Mech. Eng. 15(2) (2015) 317- 323. https://doi.org/10.1016/j.acme.2014.12.003

[3] A. Britto, R Sagai Francis, Edwin Raj and M. Carolin Mobel, Prediction of shear and tensile strength of the diffusion bonded AA5083 and AA7075 aluminium alloy using ANN, Mater. Sci. Eng., A 692 (2017) 1-8. https://doi.org/10.1016/j.msea.2017.03.056

[4] S.Saravanan, K. Raghukandan, Thermal kinetics in explosive cadding of dissimilar metals, Sci. Technol. Weld. Joining. 17(2) (2012) 99-103. https://doi.org/10.1179/1362171811y.0000000080

[5] M. Acarer, B. Gulenc, F. Findik, Investigation of explosive welding parameters and their effects on micrhardness and shear strength, Mater. Des. 24 980 (2003) 659-664. https://doi.org/10.1016/s0261-3069(03)00066-9

[6] K. Raghukandan, Analysis of the explosive cladding of cu-low carbon steel plates, J.Mater. Process. Technol. 139 (2003) 573-577. https://doi.org/10.1016/s0924-0136(03)00539-9

[7] S.A.A Akbari Mousavi, A.A Dashti , A. Halvaee, Effect of Operational Parameters and Heat treatments on the Aluminium-Steel Explosively Composite Plates, Adv. Mater. Res. 264-265 (2011) 223-228. https://doi.org/10.4028/www.scientific.net/amr.264-265.223

[8] S.Saravanan, K Raghukandan, Influence of interlayer in explosive cladding of dissimilar materials, J. Mater. Manuf. Process. 28(5) (2013) 589-594. https://doi.org/10.1080/10426914.2012.736665

[9] P. Tamilchelvan, K. Raghukandan , S. Saravanan, Kinetic energy dissipation in Ti-SS explosivecladding with multiloading ratios, Iran. J. Sci. Technol. Transactions of Mechanical Engineering 38 (M1) (2014) 91-96.

[10] M. Acarer, B. Demir, An investigation of mechanical and metallurgical properties of explosive welded aluminum-dual phase steel, Mater. Lett. 62(25) (2008) 4158-4160. https://doi.org/10.1016/j.matlet.2008.05.060

[11] ASTM E 384-99, Standard Test for Microindentation Hardness of Materials, ASTM international, 1999

[12] S.A.A.Mousavi, P.F Sartangi, Experimental investigation of explosive welding of cp-titanium/AISI 304 stainless steel, Mater. Des. 30(3) (2009) 459-468. https://doi.org/10.1016/j.matdes.2008.06.016

[13] E.Zanani ,G.H Liaghat, Explosive welding of stainless steel-carbon steel coaxial pipes, Mater. Sci. 47(2) (2012) 99-103. https://doi.org/10.1007/s10853-011-5841-9

[14] Somasundaram, Raghukandan Krishnmurthy, Hokamoto Kazuyuki, Effect of process parameters on microstructural and mechanical properties of Ti-SS 304L explosive cladding, J. Cen. South. Univ. 24 (2017) 1245-1251. https://doi.org/10.1007/s11771-017-3528-3

[15] Kahraman N, Gulence B,fndik A, Corrosion and mechanical-microstructural aspects of dissimilar joints if Ti-6Al-4V and Al plates, Int. J. Impact. Eng. 34(8) (2007) 1432-1432. https://doi.org/10.1016/j.ijimpeng.2006.08.003

[16] Y.B. Yan, Z.W. Zhang, W. Shen, J.H.Wang, L.K. Zhang, B.A. Cin, Microstructure and properties of magnesium AZ31B-aluminium 7075 explosively welded composite plate, Mater. Sci. Eng. A. 527 (2010) 2241-2245. https://doi.org/10.1016/j.msea.2009.12.007

[17] P. Mastananiah , G. Madhusudhan Reddy, K. Satya Prasad , C.V.S Murthy, An investigation on microstructures and mechanical properties of explosive cladding c103 niobium alloy over C263 nimonic alloy, J. Mater. Process. Technol. 214(11) (2014) 2316-2324. https://doi.org/10.1016/j.jmatprotec.2014.04.025

[18] S.A.A Akbari Mousavi, S.T.S Al-Hassani, A.G Atkins , Bond strength of explosively welded specimens, Mater. Des. 29(7) (2008) 1334-1352. https://doi.org/10.1016/j.matdes.2007.06.010

[19] V.N. Rao, M.G. Reddy, S. Nagarjun, Structure and properties of explosive clad HSLA steel with titanium, Trans. Indian. Inst. Met. 67(1) (2014) 67-77. https://doi.org/10.1007/s12666-013-0313-3

Keyword Index

Author Index

About the Editors

Dr. K. Hokamoto opened his faculty position at Kumamoto University, Japan in 1987 and he is currently Professor & Deputy Director of the Institute of Pulsed Power Science, Kumamoto University. His research interest is materials processing using explosive and other impulsive phenomena and he published about 200 journal papers mainly related to such field. He received some awards such as Academic Promotion Award by Japan Welding Society in 2016, Academic Achievement Award by Committee on Impact, The Society of Materials Science, Japan in 2018. He has been organized the ESHP series since 2003 launched from Kumamoto as to activate researches in explosion, shock-wave and high-strain-rate phenomena of various materials.

Dr. K. Raghukandan is an alumnus of Annamalai University, India and joined as a faculty in 1985. He is currently the Dean, Faculty of Engineering and Technology, Annamalai University. His research interests include Materials processing using explosives and laser welding. His research contributions are close to 275. He is the receipient of William Johnson International gold medal and Distinguished teacher awards. Many of his papers have won best paper awards. He has co ordinated various prestigious project grants.

www.ingramcontent.com/pod-product-compliance
Lightning Source LLC
Chambersburg PA
CBHW071231210326
41597CB00016B/2013